FLORIDA STATE
UNIVERSITY LIBRARIES

FEB 13 1995

TALLAHASSEE, FLORIDA

GOVERNMENT–INDUSTRY RELATIONS 7

THE GERMAN TRADITION OF
ORGANIZED CAPITALISM

GOVERNMENT–INDUSTRY RELATIONS

Editors: Maurice Wright and Stephen Wilks

Volumes within this series incorporate original research into contemporary policy issues and policy-making processes in the UK, Western Europe, the United States, and South-East Asia

ALREADY PUBLISHED

Comparative Government–Industry Relations:
Western Europe, the United States, and Japan
edited by Stephen Wilks and Maurice Wright

Government and the Chemical Industry:
A Comparative Study of Britain and
West Germany
Wyn Grant, William Paterson, and Colin Whitston

Capitalism, Culture, and Economic Regulation
edited by Leigh Hancher and Michael Moran

Hostile Brothers: Competition and Closure in
the European Electronics Industry
Alan Cawson, Kevin Morgan, Douglas Webber,
Peter Holmes, and Anne Stevens

Regulating for Competition: Government, Law, and
the Pharmaceutical Industry in the United Kingdom
and France
Leigh Hancher

Privatizing Public Enterprises:
Constitutions, the State, and Regulation
in Comparative Perspective
Cosmo Graham and Tony Prosser

The German Tradition of Organized Capitalism

Self-Government in the
Coal Industry

MARTIN F. PARNELL

CLARENDON PRESS · OXFORD
1994

Oxford University Press, Walton Street, Oxford OX2 6DP
Oxford New York Toronto
Delhi Bombay Calcutta Madras Karachi
Kuala Lumpur Singapore Hong Kong Tokyo
Nairobi Dar es Salaam Cape Town
Melbourne Auckland Madrid
and associated companies in
Berlin Ibadan

Oxford is a trade mark of Oxford University Press

Published in the United States
by Oxford University Press Inc., New York

© Martin F. Parnell 1994

All rights reserved. No part of this publication may be reproduced,
stored in a retrieval system, or transmitted, in any form or by any means,
without the prior permission in writing of Oxford University Press.
Within the UK, exceptions are allowed in respect of any fair dealing for the
purpose of research or private study, or criticism or review, as permitted
under the Copyright, Designs and Patents Act, 1988, or in the case of
reprographic reproduction in accordance with the terms of the licences
issued by the Copyright Licensing Agency. Enquiries concerning
reproduction outside these terms and in other countries should be
sent to the Rights Department, Oxford University Press,
at the address above

British Library Cataloguing in Publication Data
Data available

Library of Congress Cataloging in Publication Data
Parnell, Martin.
The German tradition of organized capitalism : self-government in
the coal industry / Martin Parnell.
—(Government-industry relations ; 7)
Includes bibliographical references and index.
1. Coal trade—Government policy—Germany (West)—History—20th
century. 2. Energy policy—Germany (West) 3. Capitalism—Germany
(West)—History—20th century. I. Title. II. Series.
HD9553.6.P37 1994 338.2'724'0943—dc20 93–39356
ISBN 0-19-827761-X

1 3 5 7 9 10 8 6 4 2

Typeset by Graphicraft Typesetters Ltd., Hong Kong
Printed in Great Britain
on acid-free paper by
Antony Rowe Ltd.
Chippenham, Wilts

*To my wife
and daughters
Amanda and Suzanne*

Preface

By the outbreak of the First World War the German coal industry not only appeared to have maintained its position as one of Germany's premier industries, it had also established such organizational precedents of centralization and co-ordination, in the form of prestigious employers' associations and mighty coal syndicates, that it had become the fulcrum of a highly developed system of organized capitalism. Part I of this study analyses the implications of this, particularly in regard to relations with the state and with labour, pursuing the argument throughout the 1920s and 1930s and tracing a fluctuating line of development which nevertheless resulted in an apparent amalgamation of state and industry. Part II reveals that despite the vastly altered 'external environment' of the post-Second World War world, patterns of co-operation and concertation between state, industry, and labour reasserted themselves, displaying a continuity with pre-war values and institutions greater than the influence of exogenous pressures and incursions by Allied occupiers and by neo-liberal ideologists of the so-called social market economy. Part III of the study focuses in more detail both on that enduring and powerful institution of industrial self-government which provides the primary source and inspiration for continuity in policy-making and on the major post-war innovation: the effective integration of organized labour. The analysis not only demonstrates the paradigmatic significance of developments in the hard-coal industry for the evolution of German capitalism as such, but also reveals a framework of values and institutions compatible with certain contemporary theories of neo-corporatism.

Much German literature hitherto has, traditionally, treated self-government (*Selbstverwaltung*) as a phenomenon within the sphere of public administration, e.g. the fields of local government and national insurance. A more recent approach, as in Ronge's case (see Chapter 6), envisages the phenomenon as a kind of non-state 'politics': a form of collective industrial action facilitating the government of certain aspects of industry. There have been few attempts to portray self-government as a phenomenon with a value system and organizational propensity which is transferable to various social organizations. Neither is there a single-volume work available in either English or German which portrays the politics of the German coal industry in its historical context. This study of the concept and institution of self-government addresses both issues.

The present approach presents *Selbstverwaltung* primarily as a politico-economic development which emerged historically in response to certain enduring features of Germany's political and industrial structure. The most permanent feature, if not foundation, of that system has been the state. This is perhaps one reason for the misunderstanding surrounding the term. 'Self-government', 'self-regulation', and 'self-organization' are expressions which suggest an autonomy often simply not obtaining. The autonomy in question is particular and specific, it is autonomy in relation to the state. This is not to imply that a state orientation in politics is uniquely German, but that *Selbstverwaltung* arose as a particular response to an already highly developed (and in Prussia militarized) form of central state government.

One example of the misunderstandings which can arise is to be found in Gillingham's otherwise sound and important work on the coal industry and politics in the Third Reich. He writes that 'a system of "industrial self-administration" emerged from "Gleichschaltung" '. Whereas, in fact, self-administration did not emerge in the coal industry during the Third Reich; it was an adaptation of pre-existing arrangements. Greater historical depth is necessary for a full analysis. Historical definition of the concept of self-government does not lend itself easily to a neat set of necessary and sufficient conditions, yet the highly organized nature of German industry requires satisfactory explanation. Werner Abelshauser's valuable neo-corporatist analysis rightly stresses the role of trade associations and the banks in the organizing process, but the how and why of the evolution of German organized capitalism lacks a unifying element which notions of (industrial) self-government can remedy.

The present study argues that organizational developments in the *montan*-sector in general and the coal industry in particular possess a crucial State dimension which can be analysed coherently. Although the German coal industry possesses a very distinctive relationship with the state, it is not sufficiently atypical as to undermine its 'model' quality. It is argued that, with appropriate qualifications, certain practices and precedents of government–industry relations in the coal industry have been and are transferable not only to other industries but possibly to other social institutions.

Such a study as this would not have been possible without direct and indirect help and assistance from a wide range of quarters. In particular, I should like to thank Professor Kenneth Dyson, who gave the initial impulse for this study, Dr Karl Koch, and, above all, Professor Stephen Wilks for his ready and invaluable advice and counsel. I have also never failed to be impressed by, and grateful for, the unstinting co-operation of German association and government officials. My thanks go particularly to Helmut Burckhardt, Chief Librarian, and his staff at the Bergbau-Bücherei in Essen; to the mining union leaders Adolf Schmidt and Heinz-Werner Meyer; and to civil servants in Bonn and Düsseldorf, most especially to Karl Hubert Coerdt, a constant and ready source of knowledge and experience. However, any possible errors of fact or judgement are the sole responsibility of the author. A final debt of gratitude is owed to my typist, Mrs Julie Ward.

M. P.

Contents

Abbreviations	xii
Introduction	1
PART I The German Hard-Coal Industry before 1945	7
1. Coal, Capitalism, and the State	9
2. The State, War, and Industry	39
PART II The West German Coal Industry, 1945–1990	63
3. Coal, Politics, and Industrial Continuity	65
4. The Rise Before the Fall, 1945–1957	80
5. Coal in Acute Crisis, 1958–1969	96
6. Twenty Years of Retrenchment, 1970–1990	129
PART III Continuities in Sectoral Self-Governance	149
7. The German Tradition of Self-Regulation	151
8. Coal and Self-Government	174
9. Coal, Corporatism, and Labour	200
10. Conclusion: Coal and Contemporary Organized Capitalism	225
Bibliography	241
Index	257

Abbreviations

APO	Ausserparlamentarische Opposition
ASV	A. Schaafhausensche Bankverein
ATH	August-Thyssen-Hütte
BDA	Bundesvereinigung der deutschen Arbeitgeberverbände
BDI	Bundesverband der Deutschen Industrie
BEV	Bergbau-Elektrizitäts-Verbundgemeinschaft
BHV	Bund der Heimatvertriebenen
CDU	Christlich-Demokratische Union
c.e.	coal equivalent
cif	cost insurance freight
DAF	Deutsche Arbeitsfront
DEA	Deutsche Erdöl AG
DGB	Deutscher Gewerkschaftsbund
DKBL	Deutsche Kohlenbergbau-Leitung
DP	Deutsche Partei
EBV	Eschweiler Bergwerksverein
ECSC	European Coal and Steel Community
GBAG	Gelsenberg AG
IAGI	Imperial Association of German Industry
IAR	International Authority of the Ruhr
ICU	Imperial Coal Union
IG	Industriegewerkschaft
IG	Interessengemeinschaft
IGBE	Industriegewerkschaft Bergbau and Energie
KAL	Knappschaftsausgleichsleistung
KGaA	Kommanditgesellschaft auf Aktien
KPD	Kommunistische Partei Deutschlands
MdB	Mitglied des Bundestages
MdL	Mitglied des Landtages
MdR	Mitglied des Reichtages
NPD	Nationaldemokratische Partei Deutschlands
NRW	North-Rhine Westphalia
OMS	output per manshift
p.e.c.	primary energy consumption
RAG	Ruhrkohle AG
RCDO	Ruhr Coal Distribution Office

ABBREVIATIONS

RWCS	Rhenish-Westphalian Coal Syndicate
RWE	Rheinisch-Westfälische Elektrizitätsgesellschaft
SDS	Sozialistische Deutsche Studenten
SPD	Sozialdemokratische Partei Deutschlands
STEAG	Steinkohle-Elektrizitäts-AG
USPD	Unabhängige Sozialdemokratische Partei Deutschlands
UVR	Unternehmerverband Ruhrbergbau
VEBA	Vereinigte Elektrizitäts- und Bergwerksgesellschaft
VIAG	Vereinigte Industieunternehmungen AS

Introduction

In May 1987 the last coal-producing pit in Dortmund, traditionally a core area of the Ruhr coalfield, closed down with the loss of over 2,000 jobs. Ten months later, in January 1988, the West German coal producers agreed in Bonn to a major pit closure programme that would cut production by approximately 18–20 per cent over an eight-year period, with redundancies totalling 30,000. The primary objective was to reduce output, by 1995, from the 1986 level of 83 million tonnes by between 13–15 million tonnes.

Had the West German coal industry suddenly been confronted with a dramatic reversal, or did this new arrangement represent just the latest stage in a long-standing, ongoing retrenchment? Such a question had become all the more topical and urgent within a problematic environment exacerbated by the sharply contested redundancies in the West German steel industry whose own crisis appeared to be deepening. These developments appeared to confirm the apparently irreversible demise of those two traditional industries which had been the twin pillars of the industrial revolution and which had contributed so profoundly to Germany's industrial pre-eminence not just on the European scene but, indeed, on the world stage.

In the case of coal, subsidies amounting to 24 billion DM since 1980 had failed to compensate for the coal industry's exchange rate vulnerability in the face of the strengthening DM. International coal prices are calculated on a dollar basis and foreign coal was available at prices of less than 100 DM per tonne, compared to West German production costs of approximately 260–90 DM per tonne. Government disinclination to continue subsidies at such a level, along with its intention to lower the 'coal penny' (a charge on electricity bills to make it possible for power producers to buy exclusively from German mines) contributed towards making the situation all the more critical. With annual subsidies having reached 10 billion DM in 1987, something drastic had to be contemplated.

Output had been steadily declining, at the very least since 1970 when total German output had stood at 111 million tonnes. Twelve years later the Statistical Office of the European Community in Brussels was still referring to the 'basic weakness' (*Grundschwäche*) of the European coal sector,[1] by far the greater proportion of whose continental output was being supplied by the Federal Republic where steadily increasing productivity was simply augmenting already substantial coal stocks. Just one year later, in early August 1983, the head of the Mine Workers Union, Adolf Schmidt, an SPD MP and by no means a scaremongering militant, had warned of a 'hotly contested autumn' ('ein heisser Herbst')[2] if further pit closures were to be imposed at the forthcoming meeting of the coal authorities (*die Kohlerunde*) at the end of September.

Close examination of the union position would reveal, however, that pit closures *per se* were not being opposed, only pit closures as a simple and exclusive measure to combat coal's travails. The union also stressed the need to restrict coal imports, for tax concessions to ease the spread of coal into the heating market, and, above all, to reduce coal stocks. They suggested that, over a five-year period, the then current seventeen short-time working days per annum should be increased to twenty, still of course on full pay. A ban on overtime working was in operation and for the second year running no new miners had been recruited. Nevertheless, substantial and drastic closures appeared to be imminent, not least because of sharply falling demand from the steel industry which had fallen by no less than 8 million tons in one year. Steel, along with power generation, was one of the twin pillars of the coal industry.

The union's constructive alternative proposals reinforced its traditionally positive role in decision-making within the coal industry and were designed to ensure the industry's very survival. More specifically, the union advocated, for example, that accelerated depreciation allowances could be employed within the public sector to assist the transition to coal-fired central-heating systems. Partial closures and further combination of pits into combined coal-mines ('a total colliery': 'ein totales Bergwerk') were also advocated, along with increased social protection for miners; all of which was intended to enable the German coal industry to survive the existing 'dry period' until coal was once again called upon at a later date to expand to meet an inevitably resuscitated demand. In the meantime it was argued that the indigenous industry offered a measure of security in an insecure world energy market.

An outside observer might be entitled to ask the question as to how much longer the 'dry period' was to last as it had appeared to have existed for nearly twenty years already, with very little evidence that things would be likely to change substantially within the foreseeable future. Scepticism could be justified with evidence coming from the electricity utilities as existing nuclear power stations moved up to full generating capacity and new stations came on stream, both having represented investment on a massive scale. It was reasonable to assume that coal-fired electricity generation would come under great pressure in the near future, with a very real threat to existing coal–electricity agreements.

But the sheer magnitude and sensitivity of the decisions involved, with profound economic and social consequences, was confirmed by the importance of the industry. Ruhrkohle AG, the Ruhr coal company, was still West Germany's twelfth largest company in terms of turnover, but in terms of *Wertschöpfung* it was in fourth place and as a tax-payer it lay fifth. Perhaps most significantly of all, it was the Federal Republic's ninth largest employer, with 134,479 employees, making a total of 157,800 German mining employees, if the two smaller fields of the Saar and Aachen were taken into consideration.[3] To this total could be added the many jobs in dependent supplier industries.

So, how were all the complex problems to be resolved, if at all? Had previous precedents established a framework for decision-making which facilitated workable policy solutions, both in the present and the near future? For most of the 1980s, the European Commission appeared to have ceased to represent a major policy constraint—in a statement on the latest proposals of the German authorities it concluded that:

the package was unlikely to distort free competition because West German coal and coke stocks were already high, deliveries to other member states were expected to fall this year and because there was no evidence to show that subsidised domestic coal prices amounted to indirect state aid to industrial users.[4]

The extent and nature of German policy autonomy in this sphere becomes even more apparent in the ensuing discussion below.

For more than a decade it has become fashionable to attempt to analyse West German economic policy-making and implementation from a 'corporatist' perspective. Such an approach has passed into the conventional wisdom of much of the leading business press and finds representative expression in such references as 'the interlocking web of interest groups which run the West German economy, and . . . the Government as well'.[5] The reference to Government is crucial, not least for the implied qualification regarding the realities rather than the mythology of the so-called social market economy. According to the Kiel economic research institute only about half the West German economy is free of state regulation.[6] While fashionable, this perspective is by no means a recent, post-war development. In the admittedly simplistic words of Professor

Gerhard Fels, president of the German Research Institute, a research body linked to the Confederation of German Industry, 'Germany has always been the land of cartels under the Nazis and in the Weimar Republic. Now we have become again what we have always been—corporatists.'[7]

It would transcend by far the scope of the present work to attempt to investigate the roots of collective action in German economic affairs back to the craft associations and guilds of the Middle Ages. Nevertheless, the present study, although not originating in neo-corporatist premises,[8] is forced by the sheer weight of evidence to concede the relevance of a neo-corporatist perspective in analysing German industrial policy-making. Although the evolution of the German hard-coal industry does demonstrate distinctive, atypical features, it will be demonstrated that decision-making processes and values in the coal industry display more in common with procedures in other industries than differences which would significantly divide them.

It will be argued strongly that the German coal industry does, in fact, represent in microcosm, even if in a somewhat extreme form, the corporatist nature of the functioning of the contemporary German politico-economic system. The present longitudinal study not only pursues the argument in the case of coal from the seventeenth century onwards, it independently confirms Markovits's views that the overall system of regulations, practices, and values which now prevail in the sphere of capital–labour relations in Germany are also likely to endure for the foreseeable future in regard to the whole politico-economic system: 'Ultimately, the heavy burden of Germany's past will remain the most reliable guarantor of stable capital/labor relations in the Federal Republic.'[9] The continuing existence of a CDU/CSU/FDP government,[10] which is ostensibly hostile to organized labour, will nevertheless sustain the values of the system and will not contribute to the ultimate detriment of either the coal industry or the whole industrial system itself.

One hundred years ago events in the coal industry and responses to them were to prove *modellhaft* for much of the remainder of German industry, and the present study will establish that in many respects this is still so in the last quarter of the twentieth century. Part I examines relevant historical precedents, primarily in the coal industry; Parts II and III then analyse in detail macro- and micro-economic crisis management regarding the industry, investigating the contribution of the main policy actors/groups within such different contexts as, for example, a changing energy market and changes in the composition of the political authorities. The unifying theme which arises spontaneously and is to become

INTRODUCTION 5

progressively more explicit is that of industrial self-regulation and self-organization.

The pre-eminence of continuity displayed by the industry's evolution will be shown to have been sustained primarily in the subjective sphere of certain enduring attitudes and values which can embrace and transcend any material 'discontinuities', however severe the latter may appear to be at any one particular time. The 1988 Bonn agreement on a programme of pit closures will be seen to have been by no means exceptional but entirely predictable.

Notes

1. See the article, 'Nur die Halden sind nicht mehr so stark gewachsen', in *Handelsblatt* (5 Oct. 1982).
2. See the article, 'Schmidt: Bei Stillegung droht ein heißer Herbst', in *Westdeutsche Allgemeine Zeitung* (5 Aug. 1983).
3. This compared with 600,000 in 1957; see the article, 'Stark und vernünftig: Der Chef der IG Bergbau sichert seinen Kumpels den goldenen Handschlag', in *Die Zeit* (26 Aug. 1983).
4. See 'Bonn gains EC Support for Coal Industry Aid', in *Financial Times* (12 Jan. 1988).
5. D. Marsh, 'In the Clutches of Corporatism', in *Financial Times* (5 Nov. 1987). Cf. C. Deubner, 'Change and Internationalization in Industry: Toward a Sectoral Interpretation of West German Politics', *International Organization*, 38 (1984), 501–35.
6. See Marsh, 'In the Clutches of Corporatism'.
7. Quoted ibid.
8. See e.g. P. C. Schmitter, 'Neo-Corporatism and the State', in E. Wyn Grant, *The Political Economy of Corporatism* (London: Macmillan, 1985).
9. A. S. Markovits, *The Politics of the West German Trade Unions* (Cambridge: Cambridge University Press, 1986), 449.
10. Ibid. 426.

PART 1

The German Hard-Coal Industry before 1945

1

Coal, Capitalism, and the State

Introduction

Fundamental to an understanding of the evolution of the German coal industry, is an appreciation of the dichotomy between ownership and control—control is normally associated with ownership, but not necessarily so.[1] Change, growth, and decline in the structure, organization, and output of the German coal industry are associated with many, sometimes dramatic, changes in control, with fewer, less pervasive, changes in ownership. Yet at no time has the full nationalization of the whole industry, through the vesting of ownership in a state corporation, been seriously contemplated by government.

The many aspects of control have been a responsibility shared primarily between the state, private owners, and managers in both state and private enterprises. Joint public/private companies have existed, particularly since the First World War, but their importance has been less than that of state–industry 'co-operation' established on an industry-wide basis. The influence of organized labour on the industry's counsels was minimal before the First World War, greater thereafter, but not as decisive as after 1945. It is also of questionable validity to refer in an unqualified way to 'the coal industry'. Before the First World War the dispersed and relatively isolated geographical location of the main coalfields ensured a certain autonomy in their development and, more importantly, the 'umbilical' relationship with the iron and steel industry was to prove very problematic.

The institutional and organizational control of coal production was initially a matter for the state, and in the history of the German coal industry there are two decisive turning-points: first, in 1851, with the reduction of state control to a bare minimum, inaugurating the rapid industrialization of Germany under the aegis of capitalist entrepreneurs; secondly, the First World War and its aftermath, during which the state emphatically reasserted its authority in the management of economic affairs generally and in the exercise of control over the coal industry in particular.

Coal before the Industrial Revolution

It seems to be generally true that on the Continent the state played a much more active part in promoting economic development than was the case in Great Britain. This involvement expressed a mercantilist philosophy associated with the prevailing absolutism which Prussia's rulers exemplified. As all Germany's major coalfields lay in Prussia the decisions made by the Prussian ruler and his administrators were decisive in determining industrial development for the whole of Germany. The Prussian executive controlled mining in two major ways: through its ownership of mining enterprises, and by its creation and application of the relevant mining law.

The beginning of the modern era in German coal-mining can be traced back to the second quarter of the eighteenth century when the Prussian state began a continuous, uninterrupted interest in the industry. Mining engineer August Heinrich Decker, at the behest of the monarch, Frederick William I, introduced several reforms. His most important proposals were for a new mining ordinance and the establishment of a local mining office to oversee its application, implemented in 1737 and 1738 respectively. Costs were to be met by a levy on the local pits. This was the first public supervisory body established for Ruhr mining, privately financed yet exercising state prerogatives. The state was thereafter in a position to undertake a consciously planned increase in output and to this end ever more regulations were issued, entailing a rise in the number of officials employed in the industry.

Nevertheless, it would be mistaken to exaggerate initial progress and in 1766 Frederick the Great felt compelled to promulgate three new mining decrees, the most important being the Revised Mining Ordinance for the Duchy of Cleve, the Principality of Mörs, and the County of Mark. These ordinances introduced the 'Direction Principle', henceforth to form the basis of the administration of Prussian coal-mining law for eighty-five years.[2]

Hitherto, coal-mining had been formally considered a quarrying activity, unlike mining for precious metals and other minerals. Mining regulations had only applied by derivation, or on an arbitrary, *ad hoc* basis; now coal-mining was fully and officially recognized as a legitimate mining activity on an equal standing with other forms of mining. It was on this basis that the state embarked on a most detailed regulation of the coal-mining industry.[3] The Direction Principle was intended to ensure both the careful exploitation of the coal-site and an ordered system of accounting;

management of the mine and budgetary control were transferred to the public mining office. This latter distributed profits and required deficits to be met (*Zubuße*); it also established the price of coal products. The state retained the right to reserve the use of any coal seams for itself. The application of mining law and the regulation of offences was the responsibility of the mining offices. The benefits from the miners' social welfare system (*Knappschaftswesen*: a kind of friendly society) were improved and the basis laid for the later application of its principles throughout the empire. Miners as a class received distinct privileges, including freedom from military service. State control of privately owned mines was virtually complete.[4]

Freiherr von Heinitz, Prussia's Mining Minister, issued the first regulations governing the training of officials, and a college for mining administrators and mining schools for the training of technical staff were also created. In 1784, his work was continued in the county of Mark by Freiherr von Stein, an office he was to hold for twenty years. He pursued transport improvements, particularly roadworks, introduced fiscal reforms, and also a unified system of accounting in all mining enterprises which depended on regular scrutiny, with the right to amendment, by the mining office. He was able to overcome the resistance of recalcitrant mine owners by effectively threatening them with expropriation. Under his regime the Direction Principle reached its peak. The entire conduct of private coal-mining enterprise was completely in the hands of the public mining administration, the mine owners (*Gewerke*) were mainly responsible for capital provision, receiving their profits or having to finance deficits.[5]

At this time, the Ruhr coalfield was generally held to be no more important than Ibbenbüren which was later to prove to be a very insignificant field—Prussia's premier coal-producing region lay in Upper Silesia. Here, the state's intentions were considerably more dynamic and innovatitive: between 1786 and 1811 three large state-owned collieries were opened. These collieries were deliberately established and run as pace-setters for the great feudal landlords to emulate. They were established on big seams, operated on a large scale, economizing on unit costs, and were kept in the forefront of technical development. Following the state's example, the Upper Silesian aristocracy provided capitalist entrepreneurs from within its own ranks, with their serfs becoming miners.

In 1806, the Ruhr was to be occupied by the French for seven years. Düsseldorf became the new centre for mining administration, but in practical terms the French presence did not materially affect either the steady growth of output nor increasing technical progress in the installing of steam-driven machinery. That growth did not increase still faster was not so much due to restrictive bureaucratic practices but to a lack of demand to justify increased output; over-production was a constant threat. French influence on mining law and administration was actually opposed to the Direction Principle.[6] The bulk of the Saar province was acquired by Prussia in 1815, all the mining companies except one were already state owned, having been acquired by the then head of state, Prince Wilhelm Heinrich (1740–68), some sixty years before. The Prussian government wasted little time in reorganizing the industry, introducing technical improvements, and promoting welfare services for the miners. The resultant expansion of the nationalized coal-mines in the Prussian part of the Saar owed much to the able manner in which they were run by Leopold Selle who was in charge of the Mining Office at Saarbrücken for over forty years (1816–57).

Technical development in German mining in the first half of the nineteenth century involved primarily the introduction and improvement of steam-driven machinery such as pumps and winding gear.[7] But progress in deep mining and related spheres was not particularly rapid because there was no incentive: demand was not sufficient to justify the expense of the investment involved. But the situation was to change dramatically, for reasons lying outside the coal industry itself. Stimuli for progress emanated in large measure from the state, but not primarily within the mining industry. The spread of the ideas of classical economists such as Adam Smith, and the example of the liberalization of French mining law, had little practical impact on the government of Prussia in the first half of the nineteenth century. Between 1826 and 1850 there had been no less than seven different major proposals to unify Prussian mining law, but they all foundered on the Direction Principle.[8]

But a new attitude had already been displayed in 1808 with the relative 'liberation' of local government by means of a municipal ordinance, greatly reducing state tutelage in local administrative affairs. In 1810, *Gewerbefreiheit* (liberty to exercise a trade) was promulgated; it made membership of the guilds voluntary. Primarily following the French example of the Société Anonyme, a law was passed on 3 November 1838 permitting joint-stock companies to be formed; further reform followed on 9 November 1843, a vital development for substantial progress in hard coal-mining, allowing the large amounts of capital required for investment in deep mining to be raised. The first joint-stock companies in the Ruhr coalfield were founded by foreigners, who were not familiar with the *Gewerkschaft* organization.

The very first was the Hardenberg Coal Mining Company based in Düsseldorf, with a share capital of 640,000 thaler, mostly of French origin registered on 14 April 1840. The second company, registered on 31 December 1845, was the Stolberg Mining Lead and Zinc Manufacturing Company. The third, registered on 21 September 1848, was the Anglo-Belgian Society for Rhenish Mining.[9] These companies provided a model for indigenous German investors to emulate as they arose to fulfil a demand created from the expansion released by state activity.

The first major impetus to general economic growth was provided by the creation of the *Zollverein* in 1834. It was an achievement of the Prussian executive, creating a framework within which the economy as a whole could prosper. The *Zollverein* was based on the Prussian tariff of 1818, for which Maassen had been responsible, a tariff described by Clapham as 'immeasurably the wisest and most scientific tariff then existing among the great powers'.[10] However, the exchange of goods and services not only requires an acceptable medium of exchange, it presupposes an effective distribution system which itself entails an efficient system of transport. It was precisely this crucial aspect of infrastructure which the state provided through the construction and maintenance of the main roads, rivers, and canals. The coming of the railways inaugurated a transport revolution over a wide area. It offered a speed and economy of transport which was unique. Whereas in the 1830s the freeing of the Dutch coal market and the creation of the *Zollverein* boosted coal production in the Ruhr, there can be little doubt that railway development boosted the demand for Ruhr coal in the 1840s, and to such an extent that the great coalfield lying to the north of the valley of the Ruhr began to be exploited.[11] No less than twenty-four new pits were opened in the Ruhr in the years 1841 to 1849. Total coal output from the three major fields of Upper Silesia, the Ruhr, and the Saar had amounted to just over 1,000,000 tons in 1820; by 1850 the total from these fields had increased to over 6,000,000 tons.[12] The potential of the energy source which fuelled the industrial revolution was finally beginning to be tapped in Germany in an inexorable manner: industrialization proper had begun.

The many initiatives taken by the Prussian state to promote industrialization are generally well documented in the literature on the period. But the state was not acting according to a balanced and co-ordinated plan, so different aspects of state activity contradicted one another and were creating a tension that needed to be resolved. An internal dynamic of economic development had been unleashed, by, for instance, the customs union and the railways, which was coming into conflict with what was proving to be a restrictive state role in the coal industry. Thus, Kemp concludes that 'The longevity of the state interest in mining and manufacturing was, in the first half of the nineteenth century at least, a sign of the belatedness of German development'.[13]

Due to careful nurture and fostering by the state, the German economy now stood on the threshold of major industrial advance. The demand for coal was growing enormously, but a freer response was required from coal production, and the three modes of state activity in the industry: the 'state as innovator' (as in Upper Silesia), the 'state as owner' (as in the Saar), and the 'state as administrator' (as in the Ruhr) were proving too restrictive, particularly in the area of finance. A serious bottleneck was impending. Where the state actively influenced economic development the impact was often positive, where that influence passed into control, it had become, at least at this period, negative—perhaps it was not a coincidence that in Upper Silesia the growth in coal output, where state control was least, was substantially higher than elsewhere.

In sixty-five years German coal output was to expand at such a rate that at the outbreak of the First World War production rivalled that of Great Britain. In 1913 Germany produced 191,500,000 metric tons of hard coal and 87,500,000 metric tons of lignite including the hard-coal yield of Saxony in the first figure and Bavarian *Pechkohle* in the second; Great Britain produced 292,000,000.[14] This expansion had come to be controlled by a highly organized system of monopoly capitalism, characterized by all-powerful cartels. However, this rise of the German coal industry evinced important changes both in its structure and the climate of opinion within which it operated.

Three fairly clear sub-phases can be described, each lasting roughly twenty to twenty-five years: the first is a period, lasting from the mid-century to the mid-1870s, in which state influence in managing both the economic affairs of the nation and of the coal industry was minimal; a second period of barely twenty years, from the mid-1870s to the mid-1890s, in which government resumed a more active role in economic management, but still largely ignored the coal industry; and a third period, approximately the twenty years preceding the First World War, when the state renewed its interest in the coal industry, leading to progressive intervention.

The first sub-phase is coincidental with the period of the maximum influence in Germany of *laissez-faire* doctrines of individualism and free trade. It is probably not possible to verify with certainty the precise extent to which this philosophy had pervaded ruling government circles, but in very practical terms the authoritarian regimes holding power in certain Continental countries in the period of reaction that followed the revolutions of 1848 relied upon the support of the middle classes and passed laws favourable to the expansion of industrial and commercial activities. In the 1850s and 1860s, substantial reform of mining law took place, with almost revolutionary impact.

There can be little doubt that the year 1851 marks the fundamental watershed in the German coal-mining industry in the nineteenth century— two significant laws were passed on 12 May. Both laws offered irresistible incentives to private entrepreneurs, one reduced the *Zehnt* (the 10 per cent tax on gross profits) to 5 per cent, but, much more importantly, a second legal reform, applying to the co-owners of the typical mining organization, the *Gewerkschaft*, awarded full commercial and technical autonomy and responsibility to them and their nominated managers. The special status of private collieries in Prussia was thereby ended, as this reform applied equally to the joint-stock companies. In 1860, a law demolished, in practice, the last remains of the Direction Principle by freeing prospecting and mining from the supervision of the mining office. An exception was maintained for safety purposes both internally, in the mine, and externally, regarding the environmental impact, particularly on public works. In effect, an Inspection Principle was introduced. The responsibility of the Prussian authorities had reverted to their traditional one of police duty (*Polizeipflicht*) alone.

The next major Act, of 1865, represented a milestone in the history of German mining law, remaining in force for 115 years. Not only did it enshrine the Inspection Principle, it presented a definitive codification of the right to mine (*Bergbaufreiheit*); it abolished the last vestiges of *Regalrecht*, the prerogatives of the nobility, and it unequivocally established the right of a prospective mine owner to state authorization. Now a unified mining law, based on a minimal state responsibility, was valid throughout Prussia.

This spate of mining legislation did not overlook the miner. In the 1850s and 1860s, new laws and regulations modernized the organization and administration of the *Knappschaft*, the miners' welfare organization. Hitherto, the privileges enjoyed by the miners had made 'them a caste apart': free from personal taxation, from call-up, and subject to a separate legal system. They enjoyed the special protection of their overlord and the benefits of the *Knappschaft*.[15] However, a Prussian law of 1860, introducing the 'free contract of employment', negotiated on an individual basis between employer and employee, and the reform of the miners' guild-like *Knappschaft*, concluded in 1863, abolished most of these privileges. They ceased to be organizations recognized and protected by the state and became purely voluntary associations, financed and operated by the owners and the workers. The new regulations increased employers' contributions to one-half. This shared self-administration was termed *Selbstverwaltung*, whereas, hitherto, since the time of von Stein, *Selbstverwaltung* had referred to the miners' own self-regulation in the *Knappschaft*, subject only to the input of one or two public officials acting as watchdogs. This development was not welcomed by the miners for they feared for their independence, fears which later proved to be justified. The status of the miner declined, and, with the free movement of labour from one part of Prussia to another, the work-force expanded rapidly and a proletarianization of the mineworker into just one manual labourer among others was hereby inaugurated.

Private enterprise had never been so free of obligations, either to the state or to the workers: economic liberalism had reached its zenith in Germany. This completely new legal framework facilitated the take-off of the German industrial economy. In Henderson's formulation:

Germany's main branches of manufacture expanded to such an extent that the whole character of the economy was changed. From being predominantly an agrarian country with industries organized on the basis of domestic crafts Germany became a great manufacturing centre with large units of production.[16]

In just twenty years, total German output of coal climbed to 29.4 million tons (1871) and with that Germany became the foremost continental producer of coal. By 1870, the Ruhr had become the premier coal-producing region, with an output of 11.8 million tons, twice that of Upper Silesia (6.5 million tons). But why did the Ruhr become the main centre of growth? The economic effects of the *Zollverein* were to draw an incipient German nation ever closer together so that, although situated on the periphery of Prussia, the Ruhr was not a remote location within this Greater Germany, especially with a transport network which was good and improving fast. The area was doubly fortunate in having access, via the rivers Ruhr and Lippe, to the Rhine, the major waterway of western Europe. The Ruhr's main asset, however, was its very ample supply of every type of coal: long-flame coals, gas coals, coking coals, semi-anthracite, and anthracite, for driving steam-engines, making gas, and smelting ores—above all, it was the huge reserves of first-rate coking coal which promoted and assured the Ruhr's ascendance.[17] The iron and steel industry established itself on the coalfield and formed a thriving industrial complex, the powerhouse of the German industrial revolution.

COAL, CAPITALISM, AND THE STATE

The railway continued to act as the motor of industrialization—the density of the German railway network between 1850 and 1866 increased rapidly; by the mid-1850s this system of Germanic lines which ended at Laibach had become 'by far the most remarkable piece of continuous railway in Europe'.[18] In 1860, in Prussia alone, 3,500 miles of railway were in existence.[19] The rapid growth in transport capability was matched by a concomitant expansion of coal output entailing the opening up of new coal-mines, particularly north of the River Ruhr. The boom of 1851–7 saw total German production more than double from 5.8 million to 14.8 million tons, but it fell back slightly to about 12,300,000 in 1860. Clapham summarizes: 'Disunited Germany had been helped by her railways and her far superior resources to get ahead of France and Belgium.'[20]

The rising joint-stock companies were the main agents of the startling expansion in economic growth, with 175 joint-stock companies set up in Prussia alone between 1850 and 1857. The vast overall investment in railway companies, mining companies, iron and steel works, rolling mills, and engineering companies was undertaken with the assistance of the new joint-stock banks. It was through finance thus made available, at this comparatively late stage compared with Britain, that permitted the development of production in the *montan*-industry of coal, iron, and steel in large-scale works (*Großbetriebe*). This precedent of large-scale production made an indelible impression on German capitalism and was to become a major influence on its development.

The very first joint-stock bank to be founded was the A. Schaafhausensche Bank Verein of Cologne, which soon established close links with the *montan*-industry which were to be maintained throughout the century. The Bank of Trade and Industry was formed in 1853 in Darmstadt with a capital of 42,750,000 thaler. This bank was more familiarly known as the Darmstädter Bank; it officially chose its full name, according to the annual report of 1853, for the following reason:

It is called upon to further large, sound enterprises by share ownership and investing external funds, and to do its utmost to share in the responsibility, arising from its clear insight into the total situation of German industry gained from its high vantage point, for directing entrepreneurial initiative and capital into appropriate channels, according to the needs of the moment.[21]

This high perception of the credit banks' role of directing investment in industry stands as a representative statement of their self-perception of their *raison d'être*. Subsequent actions corresponded well to the intention.

Inevitably, the fortunes of the heavy industry in which the banks shared control fluctuated, particularly those of coal-mining; in 1858, a slump followed a seven-year boom. The industry responded with amalgamations. Rationalization was to be furthered by the elimination of small, inefficient pits. After 1860, the number of pits fell continuously, the concentration movement had begun to take effect. In the Ruhr in 1850 there had already been eighteen mines each producing over 50,000 tons per year. By 1862, the eleven largest mines each produced over 100,000 tons per year.[22] The joint-stock banks had reacted to the crisis by changing their emphasis from capital loans to developing current credit facilities; they further responded by moving more into government stocks and developing their foreign business.

Perhaps one of the most important consequences of that first collapse in coal prices was the reaction of the coal owners: on 17 December 1858, the oldest and most important German employers' association was formed in Essen: the Association for Mining Interests in the Mining Area of Dortmund.[23] The general purpose of this mine owner's association was collectively to represent mining interests in relation to government. The first priority was to remove the last vestiges of state tutelage as embodied in the Direction Principle. Another vital concern was railway freight rates as most coal was transported by rail. The association was also particularly concerned to extend the market for coal. Hitherto, the state had restricted output when it was necessary to limit over-production. With that constraint removed and private producers reluctant to cut production because of the sizeable investment already undertaken, efforts to increase demand were the only alternative. Such activity exemplified two significant developments: production levels were not to be decided exclusively by unpredictable fluctuations in demand, and cultivation of the market was considered a joint responsibility, not primarily that of individual companies.

The first, fourteen-member executive committee included such names as Haniel, Stinnes, and Mulvany. Its first chairman was doctor of law Friedrick Hammacher who remained at the head until 1890. Long before that date he had become a member of the Prussian House of Representatives and of the German Reichstag, eventually becoming chairman of the National Liberals. Over the coming years the Mining Association was to develop closer links with that party than with any other, several Association officials becoming National Liberal MPs. Such a development was by no means unique, it was symptomatic of a characteristic feature of Wilhelmine Germany seen in 'the co-operation, indeed the complicity of interest groups with the political parties leading to the "coupling of offices" of association leadership, company management, and regional and national parliamentary representation'.[24] The mine owners provided a particularly prominent example of integration with the state. Kealey observes that 'for decades the Association for Mining Interests had practically occupied the position of a public office in state and society . . . the AMI was integrated at all levels into the bureaucratic system of the Kaiser's Empire'.[25]

The Mining Association also played a major role within two very influential owners' organizations: the Association for the Preservation of Common Interests in the Rhineland and Westphalia of which it was co-founder, and the Central Association of German Industrialists. Common motives held these industrial organizations together: first, to ensure that government was fully cognizant of the relevant industries' wishes and concerns, particularly in regard to protection as opposed to free trade; and, secondly, to organize effective resistance to encroachments by the state, the most effective way being to anticipate state demands in order to pre-empt state intervention.

The mineworkers were slower to organize and, in many respects, less successful. The roots of trade union organization for coal-miners were to be found in the *Knappschaften* whose senior members, *Knappschaftsälteste*, were the first official representatives of labour interests. From the mid-1850s onwards a further form of representation was provided by miners' associations, *Knappenvereine*, which from their inception were closely linked with the Catholic Christian-social movement. Although the primary purpose of these associations was the cultivation of tradition and promoting social intercourse among miners, political matters thrust themselves to the fore and members became involved in the industrial protest of the period. But no actual union organization acquired much influence at this time, membership was too small: it was so difficult to organize such a deeply conservative group of workers. They were very status conscious and, despite the great influx of new workers, clung to their traditions and consciousness of their own special identity, a process enhanced by the isolated nature of their settlements and the peculiar conditions of work underground.[26] Thus considerable social solidarity among miners was not translated into political and industrial organization. As the 1870s dawned there were already sharp divisions between Christian-social (Catholic), evangelical-social (Protestant), and social democratic miners.[27]

The gradual improvement in demand for, and profitability of, coal, commencing in 1865, turned into a huge boom in 1870. This second great boom period for coal, peaking in the three years 1870–3 of the *Gründer jahre* (Foundation Years), marked the end of one era and the beginning of the next. A great national boom, prompted psychologically by the boost in confidence from a successfully conducted war and materially by the huge French idemnity which was equal to five billion francs, one-third of one year of German gross national product, stimulated for once a coal demand which matched coal production. The newly amalgamated companies were able to exploit the situation by raising output substantially.[28] This symbolic outburst of national, economic energy may have marked the beginning of Germany's emergence as a major industrial power, but it was also the last occasion on which industrial activity in general, and that of heavy industry in particular, prospered in a competitive, liberal, free enterprise climate.

Coal and the New Capitalism

The next twenty years were to witness a substantial increase in the size of industrial and commercial enterprises, particularly in heavy industry. However, the increasing scale of investment required also entailed a concomitant increase in risk. One way to minimize the latter is to minimize and regularize competition. For such intentions to be realized a prevailing socio-economic climate needs to obtain in which free competition is not considered essential, and also a means of organizing and managing competition is required. The economic collapse following the intense speculation of the Foundation Years inaugurated a transformation in the attitudes of opinion leaders in government, commerce, and industry towards unregulated, uncontrolled competition. This was reflected to a great extent in increased efforts by mine owners' associations to promote the co-ordination of production and marketing and by the growing concentration of the banks.

The mid-to-late 1870s mark a transition in the German state's basic approach to societal developments generally and to economic affairs in particular; Wehler refers to 'a fundamental change in the structure of institutions and social arrangements'.[29] This trend is symbolized by the Trade Act of 1879 which marked the abandonment of the free trade doctrines of *laissez-faire* and inaugurated protection of the industrial and agricultural economy. The tariffs may not have been particularly high on an international comparison but they were, for example, sufficient to protect the German iron and steel industry from British competition.

Generally this protectionism furthered monopolistic tendencies at work in the German economy so that, for instance, industrial monopolies could safely raise prices to the tariff level. The political decision to abandon free trade was also an expression of a more activist, interventionist conception of the role of the state in directing the nation's affairs, this being confirmed by the anti-socialist laws of 1878, the social legislation of the 1880s, the subsidies for steamship companies, and, above all, the decision to nationalize the whole of Germany's railway system. In no sense was the Reich government under Bismarck a 'night-watchman'.

According to Wessels,[30] coal was harder hit by the Foundation Crisis (*Gründerkrise*) than any other industry. Rigid conditions of production, large capacities, and very high closing-down costs all prevented supply adjusting to demand. Coal prices fell to 4.71 marks per ton in 1879 and remained between five to six marks until 1886. Low prices did not always entail minimal or non-existent profits; because of the so-called 'free labour contract' wages could be lowered and some companies actually increased their profits in the lean years. Miners' wages were very depressed between 1866 and 1887. On the whole, however, prices were so low that many pits could not cover their costs, with production rising faster than demand because every company pursued its own narrowly defined interest, attempting to maintain or increase its own output in order to secure the advantages of large-scale production which reduced unit costs.

It was against this background of acute uncertainty for mine owners, induced by massive fluctuations in demand and prices, that the desire to regulate production and co-ordinate marketing on an industry-wide basis was stimulated. From 1858 onwards, with increasing intensity from the mid-1870s, it was the Mining Association which was at the centre of efforts, direct and indirect, to co-ordinate and unify employers' policy on an industry-wide basis. It was assisted in this process by the activities of the large credit banks moving in from Berlin. The 1880s, in particular, witnessed the resurgence of many ideas and attempts to unite mines into larger companies to economize on capital investment, reduce unit costs, disperse risk across many pits, and maintain continuity of employment. Such amalgamation represented a prerequisite for establishing effective cartelistic agreements. Despite the dedicated efforts of industrialists like Grillo, Mulvany, Gustav Mevissen, Waldthausen, and Robert Müser (following precedents set by Franz Haniel in the 1850s and the Stinnes family in the 1840s), real progress was slow. Nevertheless, continuity in the industry's counsels was assured by the creation of Coal Clubs

(*Kohlenklubs*), the very first in 1880 itself. They had been prompted by the Mining Association and gave the industry's leading men an informal opportunity to discuss common problems and to arrange acceptable solutions to them. Their discussions and negotiations, for example, formed the basis of the intermittent attempts to control prices in the 1880s.

However intractable the difficulties of satisfactorily producing and marketing coal may have seemed to the coal employers, almost equally problematic and perhaps symbolic of future problems in the organization of the industry were relations with the rapidly growing work-force. At the beginning of May 1889, Germany's first great mining strike occurred when 90,000 miners, 80 per cent of the total Ruhr work-force, went on strike to support their demands for a 15 per cent wage increase, abolition of overtime, and restoration of the eight-hour day. Bloody altercations occurred when the military was employed, increasing bitterness and causing a hardening of attitudes. Eventually, in mid-May, a provisional agreement, on the Kaiser's intervention, was reached between the Mining Association and the three mineworker representatives Schröder, Bunte, and Siegel. However, this final agreement was not endorsed by all workers and employers and although most workers returned to work the dispute threatened to re-emerge.[31]

Despite widespread dissatisfaction at the results on both sides, the mineworkers had definitely profited from the clash. An immediate short-term benefit was the creation of an effective union organization based on what later became known as the Old Association (*Alter Verband*);[32] it was formed on 20 October 1889 and had its origin in the strike leadership which had emerged during the dispute. In the medium term, legislative steps were taken to improve the miners' conditions. Specifically, an amendment to the Prussian Business Ordinance, on 29 July 1890, established courts to arbitrate on conflicts between employers and employees, including mining, and on 24 June 1892 the General Mining Act was amended by creating binding principles governing the employment contracts of miners. Particularly gratifying for the miners was the new right to appoint their own independent assessors to weigh the coal they had produced. One long-term implication was the renewed state concern for the industry. In several areas official government interest and miners' demands were to coincide, at the expense of the employers.

Another major strike, of 1905, was in many ways a carbon copy of the 1889 strike, not least in the way it caught the official union leadership by surprise and in the apparently indecisive nature of its outcome. However, one important outcome was an amendment in 1905 to the Mining Act which created workers' committees, the first instance in any industry of recognized workers' responsibility at works (*Betrieb*)/pit level, a first statutory opening for co-determination. In 1909, influenced by a major colliery disaster in France, the government issued a Regulation whereby miners themselves were allowed to elect their own safety representatives to oversee mine safety. They soon came to assume the responsibility for appointing the relevant workers committees, previously elected by all a particular colliery's workers. In the meantime, the coal employers had not remained passive: they had formed a Strike Insurance Association back in 1887 and in 1903 they formed a Colliery Owners Association (*Zechenverband*) to deal with labour relations on a formal basis.

However much industrial disputes may have enhanced cohesion and solidarity within individual mineworkers' organizations this did not translate into co-operation between, and integration of, the various unions, as was demonstrated in 1912 when a strike called by the Old Association was not supported by the Christian union. When the strike failed, divisions and bitterness between them were exacerbated, leaving the mineworkers' representatives, on the eve of the First World War, disunited and in disarray.

The conflict with the mineworkers did not delay plans to unify the industry on the basis of 'non-competitive' private enterprise, a form of monopoly capitalism. The Rhenish-Westphalian Coal Syndicate (RWCS) was finally established on 22 January 1893. It was responsible for no less than 86.8 per cent of the total Ruhr production of 31,975,642 tons; a mere thirty-one companies, mainly small ones, remained outside.[33] At long last, an effective and comprehensive form of market regulation had been established to eliminate loss-making and unhealthy competition. What had been the province of the state as recently as fifty years before—maintaining colliery profitability, rationing production, marketing policy, and price fixing—was now in the hands of an organization created for that specific purpose by private industry. Even monthly adjustment, formerly the responsibility of the mining office, was a syndicate task. The power of mine owners, organized collectively, had never been greater.[34] The founding of the RWCS represented the apotheosis of unfettered capitalism in Ruhr coal-mining.

Two potentially contradictory processes were in operation. As Wehler puts it: 'attempts to control the economic process overlapped with the progressing concentration movement.'[35] Kocka points to the paradox that fewer yet larger companies increase the likelihood of greater economic instability, hence the justification for an increased role for the state as 'stabilizer'.[36] Not least, the social implications partially explain 'the tendencies towards closer linking and enmeshing of the socio-economic and state spheres'.[37] Intervention was also furthered by that enduring tradition of 'enlightened Absolutism', namely 'welfare and police policy'. Indeed, the great miners' strike may well have prompted the conflicting major owners to come to an agreement on industry-wide organization. The state had renewed a detailed interest in the industry. Price levels, as well as the miners' lot, were of great concern to public and government alike with the perception that the apparently necessary reorganization of the industry might well advance best under the leadership of state-owned mines. Nationalization was also now on the agenda. Resolute action was perhaps required by the coal owners to forestall unwelcome state intervention. State influence was certainly reasserting itself; at the time it was fairly clear in which directions, but not precisely to what extent.

'Coal is King'?

The cartelization referred to above was the German expression of a whole phase of company co-operation within Continental industrialization, covering the period 1880–1910 and known simply as the 'cartel movement'. It occurred against a background of what Milward and Saul refer to as the 'self-reinforcing nature of industrial development once it has reached a certain level',[38] a particularly powerful stimulus being urbanization. The mutually reinforcing nature of industrial and urban growth inevitably stimulated coal production in Germany. The utilities created to provide electricity and gas were, from their inception, mixed concerns. Their organization and control was shared between the private *montan*-interests and local authorities. The industrial development of Germany's immediate neighbours, France, Belgium, Holland, Switzerland, and Austria, spurred it into becoming a regular coal exporter by the end of the nineteenth century, with 10–12 million tons being exported, although British steam coal was still a significant import, especially for the cities of Bremen, Hamburg, and Berlin. Coal and coke accounted for 2.8 per cent of total exports in 1878 and 4.03 per cent in 1903; by 1913 these exports held third place, after iron and steel, and machinery.

From the turn of the century until the First World War the coal industry became the centre of two major power struggles; that between the state and the coal barons, and that internal to the RWCS, between 'pure' coal interests and the great iron and steel concerns. The latter dispute centred on coke production, a very profitable business. Manufacturers made enormous profits, with net proceeds of between 3.26 marks and 4.34 marks per ton in 1900/1.[39] Most of Germany's coke was produced in the Ruhr and, with a few unimportant exceptions, coke works were attached to the coal-mines. The demand for coke was almost entirely industrial, mostly from the iron industry. There were few outsiders besides the *Hüttenzechen*, coal-mines owned by iron and steel firms, so this increased the power of the coke syndicates. The 1893 Ruhr coal syndicate was a cartel of pure mining companies; the *Hüttenzechen*, already producing 11 per cent of the syndicate's total, remained outside. By the turn of the century the 'outsiders' were producing 8 million tons, but the syndicate's output had barely increased at all. In times of recession, the iron and steel concerns simply sold their excess coal at prices which just undercut those of the syndicate, while the state mines acted identically once the demands of their regular customers had been met.

A number of compromise arrangements were attempted, in 1894, 1898, and, above all, in 1903. The 1903 agreement granted generous participation allowances so that in effect the iron foundries were unaffected by quotas and levies, obtaining the right to deliver any amounts they wanted to associated companies within their conglomerate. The 1903 agreement had represented a radical revision of the 1893 syndicate, it was a step towards a so-called fusion or trust. Legal disputes continued but the compromise of 1907 was not materially to inhibit the iron and steel combines. The syndicate simply could not reconcile the fundamental conflict of interest between horizontally organized coal-mines, which wished to raise prices in order to maximize profits, at least to a level consistent with maintaining demand and steady production levels, and the vertically integrated combines which wished to minimize coal prices as a means of minimizing fuel and raw material costs. Clapham aptly observes that relations between the metallurgical industries and the RWCS 'necessitated perpetual negotiation and occasional war'.[40]

There had been little public outcry against cartels leading to the sort of anti-trust legislation developed in the United States.[41] In legal judgments[42] cartel arrangements were recognized as legitimate 'self-defence' measures to ward off unacceptable developments induced by destructive competition. By the beginning of the twentieth century: 'Over the main fields of advanced industry cartels and combines had practically extinguished competition and there was a closer relationship between industrial firms and banking institutions.'[43]

The large credit bank with closest links with heavy industry in the Ruhr was the A. Schaafhausensche Bankverein (ASV), a joint-stock bank created in 1848 as the result of a transformation of a private bank which had got into serious difficulties. The continuity of the ASV's industrial activity is expressed by the fact that in 1910 one of its directors was still chairman of the supervisory board of the Hörder Mining and Iron Works Company, and the managing director of the latter was a member of the supervisory board of the ASV. Other major companies with which it was linked included Harpen Mining Company, the Hoesch Steel Works, Phoenix, and the Bochum Association. The ASV also played a significant role in the development of industrial cartels. In 1899, it established a one million mark *Syndikats-Kontor*, making it available for the administrative and representational purposes of the industrial associations and syndicates. In the same year, the ASV itself provided and performed the function of selling agency and accounting office for the Association of German Cable Manufacturers; it was similarly instrumental in creating the Steel Works Association.

The cartel movement was building up an impetus of its own and the founding of the great syndicates in the 1890s had a tremendous psychological impact from which the banks were not immune. This phenomenon was enhanced in the first decade of the Twentieth century when the big banks adopted the deliberate strategy of taking large numbers of seats on the supervisory boards of those *montan*-companies which were prominent within their respective cartels. Correspondingly, prominent industrialists, particularly from the *montan*-industry, joined the boards of the big banks. Industrialists acquired thereby additional opportunities to extend their influence and power, in regard to other firms and industrial groups and particularly within their respective cartels, and this furthered the movement towards amalgamations of all kinds. In this way cartelistic influences in industry furthered cartelization in banking, and vice versa. Although by 1912 only one big bank held more than 3 per cent of its assets in industrial stocks and shares, the Deutsche Bank was represented on the boards of no less than 159 firms. 'Their instinct was to shield from competition industrial concerns with whose interests theirs was involved.'[44] What Kocka refers to as the 'self-organisation of companies', on the basis of an 'increasing enmeshing of bank and industrial capital',[45] was reaching a very advanced stage.

Coal and the Public Power

During the period from 1900 to 1913 coal production increased by 60 million tons to 114.5 million tons.[46] This gigantic wealth-creating 'machine' was swept by a further wave of amalgamations creating huge industrial combines, conglomerates, and groupings of various kinds.[47] The whole primary sector of the economy, allied with high finance, was enmeshed in a vast interlocking system of vertical and horizontal monopolies. This represented a huge concentration of power in the hands of a relatively small number of leading industrialists, an oligarchic capitalist plutocracy. It was against this background that the state began to undertake measures to assert its authority in the national interest. Indeed, at regular intervals leading public mining officials had called for selective nationalization of coal-mines, a move supported by most political parties in the Reichstag.[48]

In 1902, a bill was introduced to parliament proposing the purchase of two collieries and one hundred standard coal lots (*Normalfelder*). The main motive for this move was military; the government had experienced difficulty with the RWCS during an international crisis provoked by events in China, as 'the fiscal mines lay to the greater part on the eastern and western borders which are endangered in the event of war'.[49] The bill granting finance was passed easily, with no opposition from private industry. Although the security argument was paramount, coal prices were also a vital issue. Ever since the founding of the RWCS prices had engaged public attention, suspicions becoming further aroused as iron works increasingly bought up pits to ensure their own independence. As a result, parliament resolved to establish a commission of inquiry, a *Kartellenquete*.[50] In coal-mining's defence Kirdorf maintained, among other things, that the industry's organized response to the recent recession had prevented a collapse on an even greater scale than that which had occurred in the 1870s. Prym concludes laconically, 'The enquiry ended with a friendly disposition towards cartels'.[51] In 1904, the Ruhr coal industry proceeded to take complete control of the coal wholesale trade and coal transport on the lower Rhine.

However, a major clash between government and private capital did occur in the pre-war years, over the proposed purchase of the Hibernia company in 1904. It was of great importance in two ways, partly because of the ambiguous role of the banks, but mainly because the state was not at that time equal to the task. The Trade Minister, Möller, had decided to purchase the relevant shares on the stock exchange secretly, using the Dresdner Bank and the ASV. The company's defence action was led by the private banking house Blechröder and the Berlin Trade Bank. The company's argument was not merely that the offer was inadequate, it also represented an attempt at 'back door' nationalization, a view strongly shared by the Mining Association. To co-ordinate defence tactics a special company was formed by the defending shareholders, under Kirdorf's leadership and popularly known as the Defiant Trust (*Trotztrust*). Its express purpose was to prevent the state acquiring further shares, and to this end it possessed a share capital of 36 million marks, half provided by the RWCS, the remainder by five big banks. Despite parliament's retrospective endorsement of the acquisition of a 46 per cent stake in the company, the government was denied a seat on company boards and was to have no voice in the company's policy decisions. Three years of litigation in the courts failed to obtain redress for the government.

Another course taken by the government to strengthen its position in the Ruhr coal-mining industry was to reserve still untapped coal lots (*Felder*) for itself. In 1905, the so-called lex Gamp was issued, a major amendment of the General Mining Act, cancelling in practice 'mining freedom' by reserving the remaining coalfields for the state.[52] Other areas of conflict also became more pronounced in the pre-war years. Taxation was one of these. Gross tax demands on mining enterprises increased by 39.7 per cent between 1908 and 1913; similarly, the state railway monopoly made constant negotiation necessary through the Mining Association. However, the main source of tension was the ever increasing amount of social legislation, and health and safety regulations, with its corollary, an increase in public officials and in their power to intervene directly in the running of collieries. It was not just the increase in costs entailed, it was the principle of interference as such which displeased.[53]

This hostility was exacerbated by the contemptuous, condescending attitude displayed by the employers towards politics, not unlike their stubborn rigidity regarding labour. The activity of parliaments and political parties was completely alien to them, it was left to the officials of the employers organizations to represent the industry in the political arena.[54] If a particular issue was important enough, the employers were sufficiently influential to turn to the leading government politicians on a direct, personal basis. The general standing of industry was distinctly higher than that of commerce and trade and the premier position of the *montan*-complex contributed towards its leaders acquiring a special sense of their own élite position (*Unternehmeraristokratie*). Their aloofness towards politics, a consciously unpolitical stance, was not compatible with the demands of a modern mass democracy and was to have ominous implications for the Weimar Republic. This fundamentally non-political philosophy was naïve, if not contradictory. A particular problem lay in the overwhelming political importance of the industrialists' own creation, the cartels. The state may have adopted in principle a position of 'benevolent neutrality' towards cartels, but this was likely to change given the ever greater accumulation of wealth and industrial power within the Ruhr which amounted to the emergence of a state within a state. The way in which the commanding heights of the economy were run was as much a political question as an economic one, particularly given the emergence of Germany as a great industrial power on the world scene.

Collaborative rather than Competitive Capitalism

Long before 1914, the coal industry had established itself as Germany's most important industry, with the Mining Association as the most prestigious and influential employers' association. The Rhenish-Westphalian Coal Syndicate was the most powerful and important cartel in the country, unparalleled in its size and comprehensive organization. It was inevitable that developments in such a key industry would have fundamental ramifications not only for the primary sector of the economy but also for the institutionalization of socio-political relationships within Germany as a whole.

The wider context had certainly not been one of *laissez-faire* liberalism, a very delicate blossom indeed, which had flowered briefly in the 1850s and 1860s, and which had withered away completely during the acute depression following the collapse of the spectacular boom of the Foundation Years. Thereafter, two major developments in the management of the economy occurred: the gradual yet irrevocable resumption of an increasingly active, interventionist role by the state, and the emergence of a collaborative form of capitalism in which doctrines of competition were distinctly subordinated to fulfilling the objectives established by the state and the major owners of capital. The market was to be seen as a device which could be manipulated, a means to an end, not an end in itself. Each boom, 1853–8 and 1870–3, had ended in a demoralizing collapse, the second more serious than the first; the possibility of a third and more shattering collapse was too daunting to contemplate.

The final creation of German unity, with the founding of the Second Empire of 1 January 1871, had not been the work of liberal democrats but the result of decisive political action by a militarily efficient state. This renewed authority of the state also confirmed the élite position of Germany's agrarian ruling class, the Junkers, who completely dominated the executive branch within the German system of government. More than any other act, the Act of 1879 which inaugurated protectionism, symbolized the practical reconciliation between industrial and agrarian capital. This alliance was state sanctioned, the expression of an industrial mercantilism which was to become increasingly more evident as Germany's industrialization progressed. The Act held a key place in Bismarck's attempt to promote economic expansion which also saw the nationalization of the railways, the extension and refinement of Germany's canal system, the expansion and improvement of public education, particularly in technical education, all measures calculated to serve the nation's dynamic economic progress. His path-breaking innovations in national social insurance were also relevant in this context.

By the turn of the century the public banking system was handling three times as much funds as the private banking sector. Between 1901 and 1913 more stocks were issued by the public sector than by trade and industry, which themselves were enjoying an unprecedented boom. Germany's large public sector was something of an exception among highly industrialized countries in the pre-war period. In Henderson's summary:

Public authorities were responsible, either entirely or in part, for the provision of communications, energy, land improvement, educational institutions and health facilities. Public undertakings included railways, collieries, ironworks, shipyards and various manufacturing enterprises. According to the industrial census of 1907 these enterprises accounted for about a tenth of all the mines, factories and transport facilities in the country. And it has been estimated that in the early years of the twentieth century between 20 and 25 per cent of investments in Germany were made through public authorities or nationalized undertakings.[55]

So Germany possessed an advanced, 'mixed' economy in which the state's role was increasing not remaining static or declining.

The growing economic importance of the state and the growth of monopoly capitalism had been more or less parallel developments in the last quarter of the nineteenth century, but in the twentieth century these lines of development began to converge, leading to a clash in the coal industry, Germany's premier industry. For a thirty-five year period, 1858–93, the Mining Association had provided an institutional framework within which coal employers could meet and discuss mutual problems on a joint basis, and where common strategies in regard to government were developed and executed. The fundamental problem of over-production was an ever present topic, promoting co-operative solutions to a situation created by the 'excessive individualism' of private companies pursuing their own individual interest through competition. The problem could have been resolved by allowing the market a free rein, with natural selection by survival of the fittest, for several companies had remained profitable throughout the long depression. But Imperial Germany's politico-economic culture was neither Darwinian nor liberal. The German state was militaristic and authoritarian, and saw the primary function of the state in the pre-eminent need to maintain order. Germany's private coal employers were also determined to establish order in the production and marketing of coal, a task to which in the 1880s the Coal Clubs devoted themselves with ever increasing intensity. This mutual concern for order established the psycho-philosophical basis of a consensus between state and capital which was to remain intact, despite serious conflict on specific issues, until the ultimate collapse of 1945.

By the turn of the century Germany had also indisputably become continental Europe's foremost industrial power. This achievement reflected the extraordinary success of capitalist enterprise, and nowhere was this industrial proficiency more concentrated and highly developed than in the coal industry of the Ruhr. But the coal industry had become more and more enmeshed with the iron and steel industry, with both industries becoming ever more closely allied with the private banks. Was this massive association of industrial and financial capital coming to resemble a state within a state, becoming in effect a rival to the authority of the state itself? Yet both capitalists and government shared the seemingly uncritical belief in the overwhelming importance of industry in general and of heavy and manufacturing industry in particular: they obviously formed the basis of the former's own power and of the state's international prowess.

Further indication of shared values and attitudes can be seen in the complete unanimity across the political spectrum in the early twentieth century regarding the desirability of limited nationalization of part of the Ruhr coal industry. A degree of nationalization was supported by every political party irrespective of the interests they represented, even top Ruhr industrialists welcomed the move. There was also little principled opposition to monopolies in general and cartels in particular. From the beginning, these forms of concentration were interpreted largely as an organizational question for the promotion of industrial efficiency, as it was believed that sufficient competition was provided on an interregional basis within Germany and/or internationally. When the Catholic Centre Party tried to raise parliamentary support for a substantial and systematic regulation of cartels it failed ignominiously, obtaining only luke-warm support from the left-wing Liberals (Progressives) and the southern National Liberals.

One important factor enhancing employer cohesion, particularly within the coal industry, was the evolving phenomenon of *Selbstverwaltung* or self-government. Employers' *Selbstverwaltung* had first been inaugurated in 1851 at company level when the private owners were granted full managerial responsibilities in their own collieries. With the formation of the Mining Association in 1858 the concept became extended to this pressure group, whose primary function was to lobby government in the industry's interest as a whole. The association gradually assumed more responsibilities as the habit of co-operation became a well-established routine. Of particular interest was its increased involvement in scientific developments and mining technology in the 1880s. It subsequently handed over control of the organization of production and marketing to the RWCS in 1893 and labour relations to the Collieries Association in 1908. All these organizations were characterized by a high level of integration and effectiveness, a testimony to a co-operative spirit which transcended particularist interests. Their ability to co-ordinate effective, industry-wide action had become well established and institutionalized.[56] It provided a leadership, a sense of direction and purpose, based on a common identity, which was unusual by Anglo-Saxon standards.

The big banks also supported solidaristic industrial behaviour. This support is conventionally related to Germany's relatively late industrialization, where the latest equipment and methods could be employed to best advantage in large-scale plants and mines.[57] The larger the productive capacity, the larger the involvement of the banks through capital investment. The greater the threat of collapse, the greater the urgency of the banks to prevent it, in their own interest. But how? This raises the second reason for monopolistic collaboration: the subjective factor of the predisposition of German businessmen to seek a co-operative rather than a competitive solution.[58] Whatever the reason, there seemed to be a predilection for order, which entailed regulation. The atomism of the market of the classical economists was the very antithesis of cohesive, centralized co-ordination. In the case of the coal industry, pressures emanating from within were absorbed, distilled, and formulated by the Mining Association in such a way as to promote concertation which in turn coincided with the interests of the banks in protecting their own investments. The most sophisticated result, the RWCS, was the epitome of organized industrial co-ordination, the result not of a negative reaction to irresistible forces, but a positive, assertive expression of an alternative mode of capitalism.

Imperial Germany, then, had developed a highly integrated form of capitalism in which collaboration and concertation clearly dominated over internal competition. Competition was by no means abolished, for example the big banks definitely competed with one another, and, within the mighty RWCS, despite fixed price levels, profits could be increased if one firm cut costs more than another: the expansion resulting from such efficiency entitling the company to an increased quota, an effective incentive. But competition was a force to be managed, controlled, and directed. This was to be achieved by co-ordination among producers, which required a co-operative attitude. The fullest expression of such collaboration on an industry-wide level was cartel agreements, which totalled 250 in Germany in 1896 and 385 in 1905, including nineteen in coal and sixty-two in the iron industry. But at company level there was a very German development which is often overlooked and which created a kind of substructure for broader collaboration, namely the *Interessengemeinschaft*, or IG (community of interest).[59]

So what kind of socio-economic system had Imperial Germany developed by the outbreak of the First World War? The whole of the heavy industry, by far the most prestigious sector of the economy, was characterized by an overlapping system of company ownership and a complex network of interlocking monopolies both horizontal (cartels) and vertical (e.g. metallurgical combines). The commanding heights of the economy represented a gigantic monopolistic octopus with tentacles which also reached into wholesale and retail trades. Employers' associations and the major credit banks had developed a highly institutionalized system of co-operation and concertation. The whole system seemed united by a strange consensus embracing the state, capital, and labour,[60] despite the evident antagonisms in the shape of the mutual antipathy of employers and employees (particularly in the coal industry), the animosity of organized labour towards the state, and the occasional, sometimes pronounced, friction of state–industry relations. The consensus rested on the rejection of the primacy of competitive values which was shared, for different reasons, by all the major socio-economic forces. The social democratic labour movement wished to abolish capitalism's competitive system of materialistic individualism. The state, the epitome of socio-political order, was increasing its role in the economy, and this entailed regulation which is the very antithesis of the totally decentralized decision-taking processes of free markets with its attendant atomism which represented anarchy to the official German mind. Similarly, the major private owners had collaborated in creating a co-operative system for concerting decision-taking in industry. In every sphere, then, co-ordination prevailed over competition: order was the highest value.

Notes

1. See H. J. Wehler, 'Der Aufstieg des Organisierten Kapitalismus und Interventionsstaates in Deutschland', in H. Berding *et al.* (eds.), *Organisierter Kapitalismus* (Göttingen: Vandenhoeck & Ruprecht, 1974), 49, and also J. Kocka, 'Organisierter Kapitalismus oder Staatsmonopolistischer Kapitalismus? Begriffliche Vorbemerkungen', ibid. 20.
2. That is, in practice; formally, they remained valid for one hundred years, until 1865.
3. Prospecting for useful minerals was unrestricted once the mining authority had given its permission.
4. German mining law was governed by two cardinal principles: *Regalrecht* and *Bergbaufreiheit*. The former expressed the fact that minerals in the subsoil, with certain listed exceptions, belonged to the king (from 1158) not the landowner; over the centuries ownership came to be transferred first to the Electors (from 1356) and then to the other princes and dignitaries of the Imperial Diet (from 1648): the *Regalherren*. These 'lords of the royal prerogative' could grant permission to prospect for minerals, assuming certain conditions were fulfilled, to any applicant, with a right to prospect even against the wishes of the landowner, i.e. the right to mine or mining freedom (*Bergbaufreiheit*).
5. Production trebled in thirty years from about 61,000 tons in the 1760s to 190,000 tons in the 1790s, total output reaching about 250,000 tons at the turn of the century.
6. See G. Gebhardt, *Ruhrbergbau: Geschichte, Aufbau und Verflechtung seiner Gesellschaften und Organisationen* (Essen: Glückauf, 1957), 6.
7. For a full account of relevant developments, see F. Schunder, *Tradition und Fortschritt: Hundert Jahre Gemeinschaftsarbeit im Ruhrbergbau* (Stuttgart: W. Kohlhammer Verlag, 1959), 79–142.
8. As early as 1809 the Prussian king (Frederick William III) had issued a Cabinet Order containing the following directive: 'But I do not wish to see private industry as restricted as hitherto in regard to running its own mines; as a rule they [state mines] should be undertaken only in quite special circumstances for the benefit of the whole.' Quoted in A. M. Prym, *Staatswirtschaft und Privatunternehmung in der Geschichte des Ruhrkohlenbergbaus* (Essen: Glückauf, 1950), 18.
9. The first such German companies were the Dahlbusch Mining Co., founded in Rotthausen in 1847; the Cologne Mining Co., founded by Gustav Mevissen in 1849 with a share capital of 2 million thaler; in 1850, Franz Haniel founded the Concordia Mining Co. in Oberhausen, with a share capital of 850,000 thaler.
10. Sir J. H. Clapham, *Economic Development of France and Germany, 1815–1914* (Cambridge: Cambridge University Press, 4th edn., 1968), 97. Its provisions

are succinctly summarized by Henderson: 'A medley of customs duties and other tolls at provincial boundaries and town gates were replaced by a single tariff levied at the frontiers of the state . . . The establishment of the Customs Union greatly facilitated the expansion of the German economy and was a factor of major importance in promoting the industrialization of the country. Hostile tariffs no longer hindered the movement of raw materials, manufactured articles and foodstuffs over a large part of Germany. Increased competition was a spur to the technical improvement and efficiency.' See W. O. Henderson, *The Industrial Revolution on the Continent* (London: Frank Cass, 1967), 15, 18.

11. As the coal-seams dipped away below the surface this exploitation required the successful deep mining techniques referred to earlier.
12. By 1850, the total Ruhr output amounted annually to 1,666,000 tons (from 388,000 tons in 1815).
13. T. Kemp, *Industrialization in Nineteenth-Century Europe* (London: Longmans, 1969), 84.
14. Clapham, *Economic Development of France and Germany*, 283.
15. e.g. free medical treatment, plus disability, widows', orphans', and death benefits.
16. Henderson, *Industrial Revolution on the Continent*, 29.
17. The success of smelting with coke first occurred in 1849 and led to the establishment of a number of blast furnaces. The comparatively late development of the metallurgical industries facilitated large-scale operations.
18. Clapham, *Economic Development of France and Germany*, 155.
19. For further details, see ibid.
20. Ibid. 281.
21. Quoted in J. Riesser, *Die deutschen Großbanken und ihre Konzentration* (Jena: Verlag von Gustav Fischer, 1910), 39.
22. As far as the individual miner was concerned, it is doubtful whether this period of consolidation much improved his economic condition: in 1869, Friedrich Engels published a report on Saxon miners working 12-hour shifts for average earnings of between two and three thalers. W. O. Henderson, *Rise of German Industrial Power, 1834–1914* (London: Temple Smith, 1975), 71–2.
23. Verein für die bergbaulichen Interessen im Oberbezirksamt Dortmund in Essen.
24. Wehler, 'Der Aufstieg des Organisierten Kapitalismus', 41; later leading to 'the new kind of interaction or even the permanent alliance of organized interests and parties', ibid.
25. M. Kealey, 'Kampfstrategien der Unternehmerschaft im Ruhrbergbau seit dem Bergarbeiterstreik von 1889', in U. Borsdorf and H. Mommsen (eds.), *Glück auf, Kameraden! Die Bergarbeiter und ihre Organisationen in Deutschland* (Cologne: Bund-Verlag, 1979), 182.

26. See W. Köllmann, 'Vom Knappen zum Bergarbeiter: Die Entstehung der Bergarbeiterschaft an der Ruhr', in Borsdorf and Mommsen (eds.), *Glück auf, Kameraden!*, 35.
27. See also H. Imbusch, *Arbeitsverhältnis und Arbeiterorganisationen im Deutschen Bergbau* (Berlin: J. H. W. Dietz, 1980; originally published by Verlag des Gewerkvereins christlicher Bergarbeiter Essen-Ruhr, 1908).
28. Many *Gewerkschaften* were changed into joint-stock companies; it was also a period with a new wave of founding of joint-stock banks, e.g. the Deutsche Bank in 1870 and the Dresdner in 1872.
29. Wehler, 'Der Aufstieg des Organisierten Kapitalismus', 40.
30. T. Wessels, 'Wirtschaftliche Probleme des Steinkohlenbergbaus in den letzten hundert Jahren', *Glückauf*, 95: 14 (1959), 895.
31. In essence, most of the workers' demands had been met, but the question of overtime and the means of regulating it by joint committees remained a bitter bone of contention.
32. Verband zur Wahrung und Förderung der bergmännischen Interessen in Rheinland und Westfalen; it was soon to be renamed the Verband der deutschen Bergleute.
33. The plan had been formulated Emil Krabler, of the Cologne Mining Company, and taken over and executed by Emil Kirdorf of the Gelsenkirchen Mining Company, assisted primarily by Robert Müser of the Harpen Mining Company and Anton Unckell, managing director of the Dortmund Coal Sales Association.
34. For details of forerunners of the RWCS in the Ruhr and other areas, and for other products, see F. Walker, 'Monopolistic Combinations in the German Coal Industry', *Publications of the American Economic Association*, 3rd ser. 5: 3 (1904).
35. Wehler, 'Der Aufstieg des Organisierten Kapitalismus', 40.
36. Kocka, 'Organisierter Kapitalismus oder Staatsmonopolistischer Kapitalismus?', 22: 'stabilising interventions by state organs, influencing growth'.
37. Continuing: 'Increasing state intervention in the economic and social sphere...', ibid. 21.
38. A. Milward and S. B. Saul, *The Development of the Economies of Continental Europe, 1850-1914* (London: George Allen & Unwin, 1977), 41.
39. Walker, 'Monopolistic Combinations', 267.
40. Clapham, *Economic Development of France and Germany*, 312. And 'such an agreement was bound to be shaky and by 1914 it had broken down completely', Milward and Saul, *Economics of Continental Europe*, 51-2.
41. Nor was the term 'monopoly' widely employed in political and economic discourse; it did not possess negative connotations.
42. Walker, 'Monopolistic Combinations', 305.
43. Kemp, *Industrialization in Europe*, 114.
44. Clapham, *Economic Development of France and Germany*, 394.

45. Kocka, 'Organisierter Kapitalismus oder Staatsmonopolistischer Kapitalismus?', 20.
46. Whereas in 1898 the largest pit produced 2,900 tons per day, by 1913 it was 5,500.
47. In an extreme case, a *Verbundgemeinschaft* could be responsible for mining coal, producing coke and other products like tar, benzene, and ammonia, linking iron works, steel-rolling mills, semi-finished stages, and engineering works, and supplying gas and electricity to industry, commerce, and domestic consumers.
48. For a detailed account of the views of the individual parties both on nationalization and cartel regulation, see M. Droste, 'Die Stellung des Ruhrbergbaus in Staat und Gesellschaft bis zum Jahre 1918', Ph.D. thesis, University of Göttingen, 1953.
49. Prym, *Staatswirtschaft und Privatunternehmung*, 33.
50. Its deliberations were later known as the 'contradictory negotiations': *Verhandlungen, Kontradiktorische, über deutsche Kartelle*—Die vom Reichsamt des Innern angestellten Erhebungen über das inländische Kartellwesen in Protokollen u. stenographischen Berichten (Bd. 1–4, Bln. 1903/05).
51. Prym, *Staatswirtschaft und Privatunternehmung*, 59.
52. A two-year period of grace witnessed the last great expansion of private prospecting extending well to the north and far to the west, on to the left bank of the Rhine as far as the Dutch border. The relevant consortium acquired no less than 275 standard lots, the primary objective being to prevent the state acquiring these coal assets.
53. e.g. the Mining Association was not able to prevent 'the impact on the works and the interference in details spreading and the free activity of works management being more and more restricted'. See Prym, *Staatswirtschaft und Privatunternehmung*, 35.
54. With them often becoming, as we have seen, National Liberal MPs.
55. Henderson, *Rise of German Industrial Power*, 177.
56. Wehler refers to the creation at that time of an 'economy structured by associations' (Wehler, *Der Aufstieg des Organisierten Kapitalismus*', 177).
57. Often associated with over-production, an endemic coal problem.
58. H. Levy, *Industrial Germany: A Study of its Monopoly Organisations and their Control by the State* (London: Frank Cass, 1966; orig. published, 1935), 4: 'The German producer was certainly not hampered in his decisions by any doctrine of laissez-faire or economic liberalism; . . . there can be no doubt German manufacturers are by their very nature in some sort of sympathy with a system of mutual consent . . . there has always been latent in the German manufacturer the co-operative (guild) spirit and also his military education leading to a certain willingness to subordinate himself, which has in many cases facilitated the formation of industrial combination' (111). '. . . the German people, through being permeated with admiration for administration

and associative organisation' (226). Cf. H. Marshall, *Industry and Trade* (London: Macmillan, 1923) Bk. I. 128, 130; Bk. II. 545, 546, and 850–1. See also Wehler, *Der Aufstieg des Organisierten Kapitalismus*', 49. This factor can, nevertheless, be overrated; there was a learning period of twenty years, 1873–93, before coal owners fully came to realize that the pursuit of self-interest was not necessarily best served by competition and that a form of co-operation might prove more satisfactory (a conclusion shared by certain leading academics of the time such as Schmoller, Schäffle, and Sombart).
59. See Levy's definition of the IG, Levy, *Industrial Germany*, 182.
60. Wehler also refers to the 'self-government' of labour and the 'gradually increasing institutional regulation of social conflicts which formally changed their character' (Wehler, 'Der Aufstieg des Organisierten Kapitalismus', 44). Cf. Abelshauser, for whom, by the end of the nineteenth century, 'the framework was established which could easily integrate the working class into the associational pattern of the German war economy and into the corporatist mode of interest politics of the Weimar Republic' (W. Abelshauser, 'The First Post-Liberal Nation: Stages in the Development of Modern Corporatism in Germany', *European History Quarterly*, 14 (1984), 291–2).

2

The State, War, and Industry

The First World War

Increasing governmental control of natural resources during the First World War appeared to be necessary to ensure efficient military mobilization.[1] This phenomenon may have represented a major innovation for Great Britain, but for Germany it was merely an intensification of a situation already manifest before the war.[2] As Abelshauser rightly points out, this development was a matter of degree rather than a fundamental change of kind:

> The German economy was already well suited in principle to the organization of war-induced planning and the direct control of production and prices within a framework of cartels extending over nearly the whole of manufacturing and heavy industry. The government used existing cartels, syndicates, trade associations of raw materials, the granting of contracts, the fixing of prices, and the control of exports and imports.[3]

To a certain extent government intervention became self-perpetuating, as one set of government measures created a situation which could only be remedied by further government actions. This is amply illustrated in the German coal industry. During the war total production fell by more than 30 per cent mainly because the most able-bodied miners were withdrawn to fight and between 70 and 80 per cent of pit supervisors, *Steiger*, were called up. An additional handicap arose from state control of the railway system which gave priority to the movement of troops and armaments, thus hindering the efficient distribution of such coal as was produced. Already, in 1914, production was down 29 million tons from 1913, and, by 1915, it was 44 million tons down. From April 1915 permission could be obtained to recall miners in special cases and there was increasing recruitment of foreign workers and employment of prisoners of war in the mines. As early as February 1915 there were 1,600 prisoners, who were skilled miners, at work in the German pits; by December they were augmented by 38,000 unskilled ones. In May 1915 Russian and Polish workers began to be employed, the number rising to 10,000 by November

1915. In the years 1915–17 there was also considerable use of 'unemployed' Belgian miners. Given the natural conflict of interest between these miners and the German miners and the deteriorating food supply situation (rationing was introduced in 1915, accompanied by the inevitable black market), it became unavoidable that output per manshift (OMS) sank, making it extremely difficult to maintain production levels.

The foremost organ of the war economy, the raw materials department of the Ministry of War,[4] had been created on 8 August 1914 and it was authorized to sequester domestically produced raw materials. On the same day, the German Industry War Committee was formed to oversee the fulfilment of industrial requirements and to provide relevant information to companies and the authorities. For a time private industry could decide itself how it would meet the demands from military and civil authorities. However, on 12 July 1915, a regulation was issued, authorized by the National Defence Act of 4 August 1914, empowering the regional authorities to establish compulsory coal syndicates if appropriate syndicates had not been formed by private companies of 'their own free will' by 15 September 1915. The regulation required that this 'free' association should involve companies accounting for at least 97 per cent of production. As the state itself produced more than 4 per cent of total coal output, private companies were under great pressure to agree on a syndicate in their own interest, in preference to a state-organized scheme, although the size of the state coal-holding ensured that the government in any case possessed substantial influence. An agreement was duly reached, giving the state special privileges, including a seat on the supervisory board, whereby the life of the RWCS was extended to 1 January 1916.

The state enforced the establishment of coal cartels because the RWCS agreement had been due to expire in December 1915 and it was most unlikely that it would have been renewed; the conflict of interest between 'pure' mines and 'tied' mines had apparently become irreconcilable. However, the syndicate's organization had become indispensable to the government in its regulation of coal production and distribution. The dismantling of such an efficient production and marketing system would have impaired the war effort. The syndicate agreement was extended to 31 March 1917,[5] with the 'only' additional condition required by the state for its continuing membership being finalization of the Hibernia purchase. This duly ensued, with the state becoming the largest single coal producer in the Ruhr, accounting for 11 per cent of a total output of 110 million tons—the state had become the most powerful influence within the syndicate. The Hibernia deal symbolized a qualified reconciliation between the state and private coal owners, significantly on the former's terms. The last wartime renewal of the contract was on 31 March 1917, for five years until 31 March 1922.[6]

In 1916 the newly created War Office established coal equalization centres to augment the work of the eleven hard- and soft-coal syndicates, but they remained ineffective. By 1917 the working of the syndicate system was failing to satisfy the state authorities; for example, despite the necessary restrictions on coal exports, some coal producers had been trying to exploit the situation of a seller's market. On 28 February 1917 therefore an Imperial Commissar for Coal Distribution was appointed with the power to issue binding directives covering virtually every aspect of the production and distribution of lignite, bituminous coal, coke, and briquettes. A special court was also established to adjudicate on disputes arising from interference with purchase contracts, and a new coal tax was introduced on 1 August 1917. State control was, in effect, total. Its authority was particularly required to enforce the rationing of coal in rural areas and small towns as the coal famine reached crisis proportions in the spring of 1918, the recently established official distribution centres playing a crucial role.

Throughout the course of the war the role of the coal employers had been progressively subordinated to the imperatives of the state, leading inevitably to demands from the Mining Association that wartime controls should not last a minute longer than necessary when the war was over. Controls, it was argued, inhibited entrepreneurial responsibility and, by generally going too far, apparently defeated their own ends. However, insult was added to injury by the sudden and dramatic increase in the status of the workers. The state, in this case represented by the military authorities, went over the heads of the employers and dealt with the workers' organizations directly. From the very beginning of the war an ever closer form of co-operation developed between the state and the workers, the cement being provided by mutual fear. The apparent lack of inhibition in this co-operation antagonized the employers, making them more hostile than ever towards the mineworkers. From the employers' point of view the high spot of their humiliation was the Law Regarding Auxiliary Service for the Fatherland of 5 December 1916. Ostensibly, the main purpose of the Act was to prevent the high turnover of labour in industry, which was allegedly undermining the war effort.[7] Both trade freedom (*Gewerbefreiheit*) and freedom of the movement of labour were abolished. Works committees of workers (the forerunners of those created permanently by Act of Parliament in 1920) were established to represent their own interests. Over a preceding period of sixty years, but particularly during the previous thirty years, German coal magnates had exemplified, perhaps more than any others, the 'master in one's own house' attitude, but now their relative position had changed drastically. But while the ascendancy of state power was permanent, the advance of labour represented in reality just one high tide in the ebb and flow of the influence of labour. However, in the immediate post-war years the employers' worst fears seemed to be confirmed, for before the year of 1918 was out they were legally required to recognize trade unions for the purposes of collective bargaining, something they had bitterly resisted for decades. In fact, they had only 'voluntarily' negotiated with the unions for the first time in October 1918, anticipating fundamental political changes after the defeat.

The Immediate Post-War Era: Coal and Socialization

Although the wartime emergency measures were considered temporary, their impact was lasting; the extent to which governments regulated economic life was greatly and permanently increased after the First World War.[8]

The post-war years up until the occupation of the Ruhr and the inflation of 1923 were characterized by substantial political, social, and economic turmoil. State authority was weakened by many factors. The Treaty of Versailles imposed humiliating conditions on the new democratic government; the parties assuming government were divided and inexperienced, yet had to rely largely on the unchanged bureaucratic personnel within the executive; the armed forces adopted a 'neutral' stance within society, becoming practically a state within a state; and the unchanged judiciary was unresponsive to the new social forces represented by the rise of an articulate and forceful labour movement. The lack of purposive direction at the centre of government entailed a weakening of state power and hindered decisive, co-ordinated action to combat the severe economic and social problems of a nation attempting to recover from the effects of the wartime blockade, of the inflation unleashed by the unsound credit policy adopted to finance the war effort, and of demobilization and reparations payments. In this context, it is understandable why certain wartime controls continued over into peacetime: they represented a framework of economic order in a potentially anarchic situation. Widespread public controls represented continuity and a common element between the authoritarian regime of the *Kaiserreich* and the demands of socialists for a new political and economic beginning. State control was also a counterweight to the destabilizing influence of socialist agitation which was itself divided between the constitutionalists of the SPD and the supporters of direct democracy through workers' and soldiers' councils led by the Spartacists. The consolidation of the SPD constitutionalists ensured that proposed reform would stem from parliamentary initiative, itself the result of consensus-building within the social democratic movement.

As early as 13 November 1918 the Prussian government had called for the socialization of factories (*Vergesellschaftung der Betriebe*). At the same time, the Reich government established a socialization commission which was due to report back in January 1919. The socialization commission, consisting of politicians and economists, issued an interim report on reorganizing the coal industry on 15 February 1919. The majority report recommended establishing an independent industrial organization; this association (*Kohlegemeinschaft*) would incorporate both state and private mines on the basis of a form of public, but not state, property which would have entailed appropriate compensation for private owners. The report was rejected by the SPD government, for the MSPD had had no genuine interest in the Commission,[9] and instead Reich plenipotentiaries were appointed for each mining region until parliament had made a permanent decision on the industry's future.

On 23 March 1919 the Socialization Act came into force. It prepared the ground for the transfer of control over the extraction of minerals to a body representative of the whole community which would be created by an appropriate Act of Parliament.[10] The Reich was authorized by law to compensate expropriated private owners. The Act expressly envisaged laws being passed to regulate on a 'commonweal' basis the exploitation of lignite, coal, coke and briquettes, hydro-electricity, and other natural sources of energy.[11] The Law Regarding the Regulation of the Coal Economy had been published the same day as the Socialization Act. The former was intended to pursue a dual purpose: to protect producers from 'uneconomic competition' and to protect consumers against discrimination in the supply of fuel and against extortionate price demands. To achieve these ends an Imperial Coal Council was established to direct the industry—it was a council of experts, initially to consist of fifty members, later to consist, in practice, of sixty members. The office of the Coal Commissar was retained, but his duties in the immediate period under review were confined primarily to matters of imports and exports, and this naturally entailed a preoccupation with reparations.

On 21 August 1919 the definitive Regulations were issued which established a four-tier system of organization for the whole industry. Of the four tiers the bottom one consisted of the compulsory regional syndicates, eleven in total. These syndicates possessed a board of management and a supervisory board upon both of which employees were represented. The main tasks of the syndicates consisted in regulating the production, 'own consumption', and the marketing of their region's output, within the limits set by the appropriate legislation. Sales emerged, however, as the primary concern. The RWCS signed its new agreement, under the new auspices, on 26 September 1919—it varied only in minor details from previous agreements.

The syndicates possessed the right to recommend price levels to the next tier, the Association of Syndicates (*Reichskohlenverband*). This second tier was an umbrella organization of the regional coal syndicates, representing the coal owners, the gas coal/coke syndicate, and the regional governments which owned their own mines. The main task of the organization was to direct the supervision of the regulation of production, own consumption, and marketing carried out by the individual syndicates. This comprehensive and centralized organization of private owners was to emerge as the effective centre of power within the structure of the industry. Nominally, the principal organ was the Reich Coal Council, which was a kind of coal parliament at the third tier. The primary task of the Coal Council was to direct the fuel sector in the interest of the nation as a whole; accordingly, the principles of 'commonweal economy' were to apply. The work of the Council's plenary assembly concentrated on debating the major, general questions of the coal sector, commissioning expert reports, and examining and authorizing the articles of association of the Association and of the syndicates. In May 1920 the Coal Council established the Grand Committee which was to participate in the setting of prices. Thereafter, binding decisions on prices could only be taken in unison between the Association and the Committee.

The fourth and final tier of supervision was represented by the Reich Economics Minister. He could reduce prices established by the Association and, from 13 October 1923, could do so without consulting either the Council or the Association. This formidable structure of controls, which was to remain formally in force throughout the period of the Weimar Republic, induced the Mining Association to conclude that the Direction Principle had been reintroduced. But the recalled Socialization Committee declared that within the new system the pits themselves had retained their autonomy, the independence of the syndicates and their Association had only been slightly restricted, and the powers of the Coal Council were proving very slight. The system was not working in practice. The Socialization Committee had been reconstituted (following the Kapp *putsch*) because public opinion clearly felt that no real change had occurred in the functioning of the industry. Power was evidently located in the Association which was dominated by the coal employers. The Commission increased in number to twenty-five (including nine socialist theorists and four top industrialists not from the coal industry), and was empowered to make recommendations. It submitted a report on 31 July 1920 consisting, among other things, of a new draft coal bill, plus a set of guiding principles (*Leitsatze*).[12] The report envisaged the final elimination of the profit motive by excluding from a commonweal economy solution private ownership of the means of production; the Association was to be disbanded and the Council to be given more powers—in particular, it would acquire title to the new form of commonweal property.

Before committing itself finally to legislation the Reich government commissioned an inquiry from the Reich Economic Council. The subcommittee thereupon established was designated a 'conciliation commission'; it consisted of three employers' representatives (Silverberg, Stinnes, and Vögler), three employees' representatives, and an independent chairman. Its report was submitted in November—it recommended state support for the establishment of several gigantic, vertically integrated mixed metallurgical concerns (coal, iron, and steel) in private ownership. The only socialistic component of the plan was the proposal to create small 100-mark shares to permit greater access to capital and to enable the workers to share in the firms' profits. Apart from providing a favourable legislative framework, the state was only required to adopt an appropriate taxation policy in regard to the industry in order to fulfil its defence of the public or national interest.

The recommendations of the Socialization Commission were never put into practice—the bill, which should have been presented to parliament in January 1921, remained unpublished: its final form had not appealed to a single interested party.[13] As the economic situation steadily deteriorated the wish to undertake experiments waned, priority was to be given to raising production now. As the political influence of the social democrats declined, with each election reducing their parliamentary representation, the impetus for a thoroughgoing socialization on the basis of a commonweal economy was gradually lost. The rearguard action of leading Ruhr industrialists to delay further reform, beyond the provisions of the 1919 Coal Act, was successful—Hugo Stinnes, and the circle of top magnates around him, witnessed in 1921 the burial of aspiration for radical change when external circumstances came to their aid.[14] The delusion persisted for a couple of years that a socialization experiment was being pursued seriously, but, with the French occupation of the Ruhr in 1923, the departure of the SPD from office, and the peaking of inflation at the end of 1923, any real likelihood of the creation of a commonweal economy, even in only the primary sector of the economy, had disappeared for the foreseeable future.

During this period of unique disturbance in German history the German labour movement made a number of important gains politically and industrially, not least by becoming recognized as a 'social partner' (to use the terminology of the period after the Second World War), but not as one of equal status. However, in the coal industry,

> the strengthening of the organizations of mine owners provided for in the Act was hardly balanced by the degree of consumer and labour participation introduced . . . In effect, the Act gave the mine owners a comprehensive and orderly system within which to function, a certain amount of good advice and some occasional interference from on high. It left the essential protection and promotion of the national interest in their hands, organizing them into a group with its own bureaucracy and leaders.[15]

The relationship between the state and industry in general, and cartels in particular, had undergone a profound change. The state's superior authority had become established as indubitable fact, but its power was not employed against monopolistic organizations. As the reserve powers of the state to exercise certain degrees of control over industry had increased it simultaneously augmented and strengthened the cartels themselves, particularly by creating a number of compulsory ones. A compromise had emerged between the state and private industry over the degree and mode of control the state should effectively be permitted, in a context where ownership was no longer a matter for discussion. A framework and network of controls had been established in which different views of the national interest had been reconciled. What is beyond doubt is the overwhelming evidence of the expanding role of the state and the mutual interpenetration of the public and private sectors.[16]

The Lull between the Storms, 1924–1929

'. . . it is easiest to look upon Germany in this period as if it were a nation already economically at war and to pose to its economic institutions the question: how could they further the expansion of the German national economy?'[17] The failure of government to carry through fundamental changes in the organization of the coal industry was partly both cause and effect of a movement of industrial concentration which dwarfed in significance earlier waves of amalgamation. Ostensibly, this development was inspired by changes in the world market situation and, above all, by territorial changes consequent upon the Treaty of Versailles. The most significant loss for the integrated metallurgical concerns was separation from their Lorraine/Luxemburg works, whereby they lost direct access to their own iron ore supplies and were denied a vital link in a production chain which began with the mining of coal and ended with the finished metal product of an engineering works. However, the major inspiration behind this industrial concentration was the determination of leading *montan*-employers to create their own kind of planned economy in such a way as to pre-empt and obviate action by the state. Again, the most prominent industrialist to exploit this initiative was Hugo Stinnes.

By this time, German heavy industry (the *montan*-sector) was dominated by eighteen to twenty mammoth concerns. From this point on, the history of Ruhr coal-mining largely reflects the evolution of these gigantic firms, as pure coal-mining companies hardly existed any more—the huge Gelsenkirchen Mining Company, having acquired a number of *Zechenhütten*, possessed the largest quotas within the Syndicate. Industrial organization was characterized more than ever by 'communities of interest' (*Interessengemeinschaften*) both between large firms and between large concerns and smaller ones.[18] Typical examples of this fresh wave of company association were the Rhein–Elbe Union, linking the Gelsenkirchen, Deutsch-Luxemburg, and Dortmund Union companies; the Klöckner and Stumm concerns which were reconstituted in 1921; and the Hoesch Steelworks which had become associated with both the Köln-Neuessen Mining Company and the Rhein–Elbe Union in 1920. German capitalism seemed more committed to ideals of co-operation than competition, and company association limited inter-company competition; syndicates regulated prices of particular products and the Mining Association represented the industry's joint interest *vis-à-vis* the state and in the sphere of sponsoring technical development.

In addition to the negative rationalization of pit closures there began a positive rationalization movement which endured for nearly twenty years. Priority was given to underground consolidation, mechanization, standardization of equipment and working methods, the establishment of pit control points, and so on. This intense rationalization effort was initially sustained on a joint, industry-wide basis (*Gemeinschaftsarbeit*) by most private owners; they were soon augmented, however, by support from organized labour which, following American experience, believed that greater efficiency through mechanization meant higher living standards for the workers. The Reid Report of 1945[19] represents an eloquent testimony to the thoroughgoing and wide-ranging success of the rationalization movement in the German coal industry. All-round innovation was pursued with a serious professionalism and dedication which was the expression of an underlying national consensus on the pursuit of German industrial supremacy which united state, capital, and labour.

The drive for efficiency was by no means the sole responsibility of both sides of industry in general, and of heavy industry in particular. As early as 1921 the Economics Minister had advocated the establishment of an Imperial Institute of Efficiency (*Reichskuratorium für Wirtschaftlichkeit*) which was in fact founded on 10 June by private industry and the research associations (*technisch-wissenschaftliche Vereine*). The state exercised a greater influence over rationalization in the coal industry than in any other industry. The technical and scientific experts of the Coal Council were required, through their appropriate committees, to supervise rationalization measures within the industry and an additional such committee was established in 1921 with the specific brief to further technical progress in the Ruhr mining industry. In 1926, the specialist technical committees of the Mining Association formally put themselves at the disposal of the Imperial Institute in order to pursue better the tasks still outstanding. The national objective of economic supremacy based on industrial efficiency was to be achieved in partnership between the state and private industry. This unity of purpose provided the basis of a cooperation which overrode conflicts between the state and private employers on specific issues; with the departure of the social democrats from office the state did not seem to pose such a threat to industry's autonomy.

Such developments, however, could not disregard the true cost of German coal—in the second half of the 1920s two important commissions of inquiry made a detailed examination of the coal industry's cost structure and of the prospects for the immediate future. After several weeks of intensive investigation the Schmalenbach Committee reported, in 1927,[20] that the average ton of marketable coal entailed a loss of 27 pfennigs, a loss which rose to 1.41 marks per ton, not including interest charges, for the pure mining concerns. Another commission (*Enqueteausschuß*), undertaking a more detailed inquiry,[21] took more than three years to report: from 30 July 1926 to October 1929. It established that on the average ton of coal costs exceeded proceeds by 0.81 pfennigs (the situation was totally different for coke). These findings (not at this stage confirming the success of rationalization) were presented as an ominous portent for German industry as a whole, given coal's key position within the national economy.

However, discussion of German coal's situation in isolation from the coking business is totally artificial and misleading. The power struggle of the pre-war years between pure mining companies and the tied mines of the metallurgical concerns had been definitively resolved: 'it is evident that the Ruhr steel mills had gained an overwhelming majority in the various syndicate organs by 1925.'[22] What had happened in the Ruhr was 'in effect the control of a sales monopoly by an important coal consumer'.[23] The continuing and highly satisfactory profitability of the coking industry was responsible for keeping the whole industry 'in the black'. Parker unequivocally confirms that the structure and nature of the German coal pricing system was established according to the industry's own wishes.[24] Nevertheless, in one sense the writing was on the wall: pure coal itself was no longer profitable, it was being maintained by the demand for coal and coke derivatives. The incipient pincer movement of increasing substitution of coal by competing fuels and increasing production costs as the pits went deeper and coal became more difficult to mine was obscured by the dislocation incurred during the Great Slump and deferred by the autarchic policies of the Nazi regime. Iron and steel interests controlled coal-mining, which controlled the coking industry, which in turn dominated the gas and basic chemicals industries.

In 1929, at the height of its prosperity and power, this complex structure of syndicates dominated by combines must be reckoned one of the most effective monopoly systems ever to have existed independently of the political support of a totalitarian state . . .[25]

The coal industry's programmes of rationalization of deep mining and manifold and extensive diversification in coal-processing and electricity supply entailed vast and expensive investment no more than 60 per cent of which could, on average, be financed internally by depreciation allowances and cross subsidization. Resort to the major joint-stock banks was automatic and inevitable, given the well-established and intimate connections.[26] The funds of these banks, decimated by the ravages of the great inflation, were for a time augmented by American speculators who channelled their investment through the existing banking mechanism. However, the umbilical relationship between banking and industry, having proved such a source of strength in the past, was soon to turn into a chronic liability. The debility of the banks, emaciated by the great inflation, was obscured by the inflow of American money. With the Great Slump which followed the Wall Street Crash, the German private banking system was in danger of collapsing like a house of cards, taking German industry with it.[27]

The imminent collapse of the commercial banking system was, because of its sheer size, such a severe economic problem that the state was forced to intervene. Just as the banks had often employed their capital resources to further reorganization and concentration in industry in the interests of efficiency, so, ironically, the state was now in an identical position in regard to the credit banks. In a situation where 'long before the banking crisis of 1931, the government had acquired a dominant position in German banking',[28] two-thirds of the banks' total capital was transferred to the public sector and two of the main banks were compelled to amalgamate. By the end of the crisis the state had acquired 34.65 per cent of the share capital of both the Deutsche Bank and the Disconto Society, 70 per cent of the Commerz- and Privatbank, and no less than 90.67 per cent of the Dresdner Bank.

the German banking system had thus been virtually nationalized by the Republican government on the very eve of the National Socialist revolution. Indirectly, the process implied that the government had also gained control over the major part of German industry.[29]

By 1932, the proportion of national income determined by the state could well have been as high as 50 per cent.

Stolper observes that 'At its demise, the Weimar Republic bequeathed to the National Socialist state an economic system which came close to being a thoroughly developed state socialism',[30] but of a peculiarly German kind, for, as Parker concludes about the remarkable, integrated system of coal syndicates which had experienced perhaps more state intervention than any other sector: 'The outstanding feature of this organization was not its effect on fuel distribution, or the ideal of national strength which it seems fitted to serve, but rather that it existed under private ownership.'[31] The state had initiated German industrialization, it had continually fostered its development, and, in nurturing the economy back to health, the state was to assume an unprecedented level of control. The influence of high finance, symbolizing the power of industrial capitalism, had been broken in Germany by the inflation and the Slump. Private industry had become so interwoven with, and dependent upon, the state that it was not in a position to regenerate itself. The reconstruction of the German national economy was not to be the work of a revolutionary labour movement but of a monolithic state apparatus guided by fascist principles.

Coal and the Third Reich

The progressive increase in state control of society and the economy in Germany, which became total during the Second World War, was the expression of the totalitarian philosophy put into practice by the National Socialists. But the important point about the Nazi 'revolution' was that it was not a revolution. The totality of the Nazi regime represented simply the continuation and the culmination of a trend which had begun in the mid-1870s when rapidly industrializing Germany turned towards protection; which was further boosted during the Wilhelminian period by the determination to achieve *Weltgeltung* (a dominant position in the international sphere); and which reached an interim peak during the First World War (partly as a result of the Allied blockade). The Weimar Republic witnessed a contest for power and influence between the state and private industry, between the millstones of which representative government was crushed. The centrifugal forces of the atomistic pluralism of the parliamentary system had proved too weak to balance the centripetal forces of the state–private industry consensus. German industrial society seemed to have reverted to type, a new form of industrial mercantilism had emerged on a higher and more sophisticated level. In retrospect the *laissez-faire* period of the 1850s and 1860s seems to have been something of an aberration, the inherent anarchy of the market seemed to be anathema to German capitalists who wished to organize and regulate as much as possible. But co-operate and co-ordinate as they might neither cartels, employers associations, nor chambers of commerce could prevent the economic collapse at the beginning of the 1930s: the only force which could guarantee order and which possessed the resources and the will to reconstruct an industrial society which had collapsed was the state.

National Socialism was not a rigid ideology and despite the commitment to planning and direction (*Lenkung*) economic measures, particularly in the first three years of power, were introduced in a piecemeal, pragmatic fashion. One of the first steps was the dismantling of the Reich Coal Council on 21 April 1933 and the removal of all employees' representatives from the Association of Syndicates and from the individual syndicates themselves. In May 1933 the trade unions were abolished and by the end of the year so too were the employers' associations, as in other spheres the leadership principle dominated. The Mining Association was saved on the intervention of Schacht, on the grounds of the Association's technical preoccupations. The organs of self-regulation (*Selbstverwaltung*) acquired a state-authorized independence; within a state-defined framework they were autonomous.[32] The only organization permitted to represent mining interests was, according to the Regulation of 17 December 1934, the Mining Economy Group in Berlin. The Ruhr mining industry belonged to this through its membership of the specialized group (*Fachgruppe*) for the hard-coal industry which embraced regional groups, the one for the Ruhr being based in Essen. Membership was compulsory. Private owners themselves possessed an advisory function only.[33] In February 1935 the official mining administration became the responsibility of the Reich,[34] under the Economics Ministry.

On the eve of the Second World War Germany's autarchic economic programme had contributed towards its national coal industry having become the largest European producer. With lignite included, German output in 1938 was 252 million tons against Britain's 232 million tons. In 1937 Britain and Germany had agreed on a division of the world coal-market. Despite the frequent organizational changes which had apparently facilitated this resurgence of the German coal industry, the war period witnessed further important changes. The final coal organization to be created during the war years was the Imperial Coal Union (ICU). Formed on the dissolution of the office of Coal Commissar this latest coal body represented the last of a series of compromises between the state and the coal industry over the creation and implementation of coal policy, all the more critical in wartime. The autonomy of the industry's own organization, the ICU, was more apparent than real. The senior management of the organization was in the hands of the chairman and an advisory presidium both of which were appointed by the state according to the leader principle. This chairman was the Economics Minister's agent and he was also responsible to the Plenipotentiary for the Four Year Plan for coal's fulfilment of its obligations under the Plan. He was renamed the Imperial Coal Agent and possessed his own administrative department with the status of an Imperial Office (*Reichsstelle*). As both Agent and presidium were appointed by the Planning Plenipotentiary and had to follow government directives, the Union was directly subordinated to the state.

Coal production was maintained until the end of 1943 but by 1944 the Allied air raids began to take their toll. On 2 September 1943, the Führer issued his Decree of the Führer on the Concentration of the War Economy. The Coal Union was allocated to the Raw Materials Office, the Imperial Coal Office became a branch of the Planning Office, and the coal industry's administration became part of the executive branch of government. State control of the coal industry thus became total under National Socialism, *Selbstverwaltung* was, in truth, non-existent.

Conclusion

During the Wilhelminian epoch the German economy developed distinctive features: besides possessing important instances of natural resources geographically concentrated, and an uninterrupted protection by tariffs, German industry was concentrating its efficiency on heavy production and on goods of cheaper grade, highly adapted to bulk and standardized supply. On the other hand, its special and early success in applying scientific research to industry gave it a monopolist position in certain high grade branches of industry.[35]

These developments encouraged concentration and facilitated cartelization. It was in the extractive industries particularly that the apparent advantages of cartelization were demonstrated to other industries. The coal industry exemplifies how the cartel movement was partly facilitated by the movement towards concentration of units of production and enterprises, and was partly a means of furthering such concentration. The First World War also witnessed the bringing together of manufacturers from individual branches of industry under the auspices of such semi-public corporations as the War Raw Materials Associations. Leaders of cartels worked with other industrialists, spreading the idea of mutual understanding to branches hitherto free of cartelization. The habits of co-operation were to last well into the peace.

By 1925, the Imperial Association of German Industry (IAGI) estimated that there were 1,500 cartels in operation in Germany; a government survey revealed no less than 3,000, of which 2,500 were in industry. Cartels were the forerunners of industrial amalgamation. The large metallurgical combines which had come to dominate the hard-coal industry demonstrate how most big concerns were closely linked up with a great number of cartels. The supreme embodiment of this development was the creation of the United Steel Works (*Stahlverein*) in 1926. The

Stahlverein was an extreme example of an amalgamation which was a monopoly in every sense, a large concern which dominated all its fields of production. It was a huge vertical combine and was active in numerous cartels, and as such it was a monumental instance of co-ordinated, institutionalized production at the heart of Germany's industrial system.[36]

By 31 October 1927, no less than 2,106 of Germany's 12,008 joint-stock companies were incorporated in *Konzerne*. These tied concerns accounted for 60 per cent of total share capital, 11 billion out of 18 billion marks in value.[37] Both cartels and trusts were formed with the object of eliminating competition, and the holding company form of trust (like, so many examples, copied from American experience) was particularly favoured for fiscal reasons. Tax savings were to be made by forming a limited company with a small amount of capital, taking over the majority if not all the shares of companies to be amalgamated, thus easily avoiding the costly act of creating a new corporation.

Civil servants and private industrialists shared a predilection for organized control. How is this tolerance, even approval, by the state of monopoly control of industry to be explained, assuming that the aim of industrial combination of whatever form is to restrict, if not eliminate, competition, in order to increase profits? After all, it was the general view by the mid-1920s that cartels were maintaining prices above levels that would otherwise be the case. The answer is to be found in the field of industrial efficiency.

In theory, the limitation, even abolition, of competition means weakening or removing the impetus to innovate because the price mechanism of free markets cannot function and direct demand to where supply is most cost effective. Without the spur of competition, at least to survive, if not to expand, industry stagnates, becoming inefficient and backward. However, the German example offers a refutation of the belief that this must necessarily be so. Germany's success as a major trading nation was based on productive efficiency. From the very beginning, company amalgamations, whether inspired from within, resulting from changes in demand or production techniques, or from without, for example through the initiative of particular banks, had been justified as offering opportunities for rationalizing production. By concentrating and extending units of production unit costs could be reduced through large-scale production.

Although the advantages were, no doubt, often exaggerated, or a pretext for other motives, the results speak for themselves. Walker's thorough study of the first ten years of the RWCS establishes that it would be preposterous to assert that cartelization had restricted innovation and

THE STATE, WAR, AND INDUSTRY 55

limited efficiency.[38] In addition coal profits were moderate, although coke profits were exorbitant. The practical desire for industrial stability to provide a secure base for expensive and risky investment coincided with the German predilection for order and regulation, and the RWCS was its institutional apotheosis. The remarkable increase in productivity in German coal-mining in the 1920s and 1930s was the direct result of rationalization and mechanization. In fact, a social cartel of interests seemed united in a drive for ever greater efficiency and productivity which was much inspired by the exaction of reparations and the much-admired example of American technical progress. Within this context, Levy stresses the complementary rather than contradictory aspect of various monopoly organizations:

as regards the question of greater rationalization and standardization and other problems of economic and technical unification the interest of the cartels are certainly deriving advantage from the trust movement within their borders . . . elimination of uncontrolled competition coupled with schemes for better organization by the collective reducing of costs seemed to have become the acknowledged function of combination.[39]

Establishing the basis for long-term profitability was in the self-interest of industrialists and coincided with the state's desire for international prowess.

The vital and explicit necessity to increase exports in the national interest boosted the community of interest between state and industry, leading to their partial merging, and blurring in a way unparalleled in other major industrial nations the distinction between the spheres of politics and industry. An overlapping and intermeshing developed which could prove politically dangerous given the weakness of parliamentary government. Not only was the state the largest single coal producer within the industry, with a corresponding position within the coal syndicates, it acquired substantial control over the whole industry through the elaborate mechanisms of the 1919 Coal Act. The cartel legislation of 1923 gave the government substantial influence over cartelization throughout the economy. State enterprises, for example the coal-mines, operated along commercial lines, in practice losing much of their distinctive identity. The rise of the huge gas and electricity utilities, jointly owned by public authorities and private shareholders, created companies which were amongst the largest in continental Europe. The public banking system, increasingly lending to private industry, went from strength to strength.

The public banks were able to unite their resources in the Deutsche Girozentrale. The Reichsbank came to be recognized as an important instrument in regulating the business cycle. The growth and diversity of public enterprise, entailing an entangling of state and capitalist interests in a complex apparently low in conflict potential, was epitomized by the creation of VEBA and the VIAG.[40]

By the early 1930s Germany had developed an alternative form of capitalism. The Anglo-Saxon model (however much in practice the reality seemed very different) was still supposed to be a private enterprise system based on free competition, with monopolies the exception and where the state played a minimal role. In Germany, the private enterprise system was based more on co-operation and co-ordination, where free competition seemed to be the exception, and with the state playing a very substantial role. As Levy points out, 'German cartel organization comes in many cases very near the conception of a general organization of industry based upon principles of common action and associative agreement',[41] while Naphtali shows that 'among the various peoples of western culture it was precisely the German people which experienced the strongest subordination of the private economy to the general interest, in both earliest and recent times, and also, indeed, precisely in the form of the public enterprise'.[42] The concertation of state economic activity and private capitalism had attained such a level that there emerged during the Weimar Republic a deep and widespread interpenetration of the political and economic spheres, a progressive integration which led under the Nazis to a fusion of the two during the Second World War. This development has been traced in detail above by pursuing the fate of the coal industry.

However, two key issues require further elucidation: first, precisely how the state was able to proceed so smoothly to complete dominance of the economy and society; and secondly, the nature of the reaction of the German labour movement to the inter-war experience. To approach the first, it is necessary to appreciate the status and importance of the industrial organizations and their personnel created, at the very latest, by the early 1930s:

by the interlocking process of industrial organization of today leaders of one industry have become absorbed in the interests of a great many branches, and ... [such] leaders are in general important factors in the central organizations of industries and likely to be considered as bearing a good deal of public responsibility with regard to the general welfare of national industry.[43]

If the state could ensure the loyalty of such individuals and their bureaucracies, they would be able to deliver the acquiescence and co-operation of their constituencies in implementing the intentions and plans of the state.

It is unlikely that such a phenomenon would arise in a socio-economic vacuum. Certain precedents needed to obtain, embodied in an appropriate framework of values and institutions, which would facilitate the subordination of private to public interests. As has already been indicated, the German tradition of *Selbstverwaltung* (self-government or self-administration),[44] provided just such a vehicle; in Naphtali's words:

> By self-administration, constitutional law understands the exercising of administrative authority by public-law associations which are a part of, or subordinate to, the state. Administration, a sovereign right of the state, is delegated for certain purposes to lower organs. The concept of 'economic self-administration' presupposes therefore that the leadership of the economy, not only the regulation of its external conditions, is a task of the state. Proceeding from the recognition of this superior authority, then economic self-administration bodies are accordingly associations of economic units (companies) or of people active in a branch of the economy for the purpose of exercising the authority of economic leadership within the framework of an objective set by the whole of society, represented by the state.[45]

According to this analysis economic *Selbstverwaltung* represents a devolved responsibility, not an inherent right; therefore self-government is a revocable competence or responsibility, existing by the grace of the state.[46] The starting-point or basic assumption is not the inherent autonomy of industrial decision-making but the state as the primary foundation and superordinate institution in regard to all social organizations. Who owns the means of production is a question which cannot alter the basic social and constitutional principle as long as the appropriate consensus prevails.

The second key issue concerns where organized labour stood in relation to the co-operation between, and increasing integration of, state and private industry. Experience in the coal industry had not encouraged optimism:

> In spite of all the newly created bodies, the preponderance of control is still in the hands of the cartels, which in general refrain from any genuine policy of 'common interest', while the admission of workmen's representatives has not been able to change the character of the cartel.[47]

Labour representatives had seemed more concerned to increase efficiency and productivity, not least to increase wages and living standards, than to create a socialist economy. A union pamphlet observed in 1926 that:

> Rationalization is necessary. In accordance with the Memorandum of the National Association we conceive rationalization, i.e. the application of all technical and organizational means, which are likely to increase the productivity of labour and machinery in industry, as the most important condition for improving prosperity.[48]

This dedication to efficiency, allied with a readiness to compromise which belied many of the radical and militant declarations of social democracy, produced an underlying socio-economic consensus upon which industrial efficiency, increased exports, and the restoration of national pride depended.

The expansion of various monopolistic forms of control, and a growing role for the state, were the two dominant trends in the German politico-economic system and were greatly stimulated by the exigencies of the First World War. Within a very short time the state necessarily advanced from the position of junior partner in relation to private industry to that of senior partner. The demands of the war also necessitated a *rapprochement* with the labour movement. The co-operation of its industrial wing, the trade unions, was obtained by permitting their participation, albeit at a subordinate level, in the decision-making bodies co-ordinating the efforts of the domestic economy. The principle of recognition was more important than the actual extent of responsibility granted.

A clear hierarchy of authority emerged to direct the industrial system during the war: the state, at the apex of which stood the military High Command; then private owners, organized in appropriate associations; and finally labour, led by active trade unionists. The social democrats were excluded from government, so their main channel of influence was an indirect one through the unions. A central issue underlying the five years of turmoil which characterized the aftermath of the war concerned the possibility of restructuring that very hierarchy. The German labour movement wanted to use the opportunities granted by the political revolution of November 1918 to raise itself to a position of equality from which political, social, and economic conflicts could be resolved in a corporatist fashion.

Weaknesses within the labour movement included a lack of clearly defined objectives and, above all, confusion over the means of attaining them, difficulties which were accentuated by the rise of a radical revolutionary alternative, initially in the form of the Spartacists, later by the KPD. This weakness, combined with the resilience of private capital which had abandoned the apparent political neutrality of the pre-war years, combined to thwart social democratic aims to democratize society and the economy.

Five years of consolidation of the German economy ensued from the stabilization of the currency in 1924, with the social democrats in opposition until 1928. This recovery evinced a largely harmonious relationship between the state and the capital, a *modus vivendi* had been established between nominally equal partners. Although substantial formal rights had been granted to labour its power was more apparent than real. However, a rationalization mania swept the German economy during the second half of the 1920s, leaving industry hopelessly over-capitalized when the great crisis struck in 1929.

This clearly exacerbated the depression, causing Germany to suffer more than any other industrial nation. It seems that whenever German capitalists were left largely to their own devices they overreached themselves, causing the recession following the boom to become doubly serious. Hence the most significant development in the politico-economic system at this time was not so much the Nazis coming to power, but the end of democratic politics in 1930 when the government began to rule by decree. What became apparent under National Socialism was that the rise to pre-eminence of the state in the direction of the economy (and control of society) during the preceding forty-five years, represented, in Freudian terms, the return to the womb on the part of private industry. The private sector had been conceived during the Napoleonic Wars, as can be seen in the Cabinet order of 1809 and the *Gewerbefreiheit* of 1810, and born at mid-century, with the rise of joint-stock companies, particularly joint-stock banks, and the abandonment of the direction principle in coal-mining. After one mere generation, the state and private industry reverted to an organic integration, along a path which was far from smooth, until the state became again all-powerful, a process which culminated during the Second World War. Nowhere is this cycle more clearly demonstrated than in the coal industry.

Notes

1. See e.g. T. Schieder, *The State and Society in Our Times* (London: Thomas Nelson, 1962), 76.

2. See H. J. Wehler, 'Der Aufstieg des Organisierten Kapitalismus und Interventionsstaates in Deutschland', in H. Berding *et al.* (eds.), *Organisierter Kapitalismus* (Göttingen: Vandenhoeck & Ruprecht, 1974), 50.
3. W. Abelshauser, 'The First Post-Liberal Nation: Stages in the Development of Modern Corporatism in Germany', *European History Quarterly*, 14 (1984), 297.
4. Cf. S. B. Clough, C. Moodie, and T. Moodie, *Economic History of Europe: Twentieth Century* (London: Macmillan, 1969), 39–40.
5. A. H. Stockder, *Regulating an Industry: The Rhenish-Westphalian Coal Syndicate, 1893–1929* (New York: Columbia University Press, 1932), 18.
6. See G. Boldt, *Staat und Bergbau* (Munich: C. H. Beck, 1950), 77–9, for details of the potash cartel of 1910.
7. But the real purpose was to 'stabilize industrial relations ... necessary to integrate labour into' the system referred to in the opening paragraph (Abelshauser, 'The First Post-Liberal Nation', 279).
8. Clough *et al.*, *Economic History of Europe*, 23.
9. See C. S. Maier, 'Coal and Economic Power in the Weimar Republic: The Effects of the Coal Crisis of 1920', in W. H. Mommsen, D. Petzina, and B. Weisbrod (eds.), *Industrielles System und Politische Entwicklung in der Weimarer Republik*, ii (Düsseldorf: Athäneum Verlag, 1974), 536; and P. Wulf, 'Regierung, Parteien, Wirtschaftsverbände und die Sozialisierung des Kohlenbergbaus, 1920–1921', ibid. 648.
10. In order to fulfil two objectives: (i) to socialize suitable commercial enterprises, and in particular to take those into common ownership (*Gemeinwirtschaft*) which extract and exploit natural resources; (ii) in the case of urgent need, to regulate the production and distribution of commodities (*Güter*) on a joint/common (*gemeinwirtschaftliche*) basis.
11. An identical approach finds expression in paragraph 2 of Article 156 of the Weimar Constitution of 11 Aug. 1919, quoted in full in Boldt, *Staat und Bergbau*, 121.
12. An option was provided from two alternatives which largely differed over the time-scale of implementation (immediately or after a lengthy transition period).
13. In fact, neither the Reich government, the Prussian Trade Ministry, nor the Old Association had ever seriously intended socialization, it was a manœuvre to placate the more radical demands of the more extreme elements; see H. Mommsen, 'Sozialpolitik im Ruhrbergbau', in Mommsen *et al.* (eds.), *Industrielles System und Politische Entwicklung*, 312.
14. Cf. Wulf, who claims that one excuse for abandoning coal socialization was that such an organization would be too easily controlled by the victorious powers (Wulf, 'Regierung, Parteien, Wirtschaftsverbände', 655).
15. N. W. Parker, 'Fuel Supply and Industrial Strength: A Study of the Conditions Governing the Output and Distribution of Ruhr Coal in the Late 1920s', Ph.D. thesis, Harvard University, 1951, 121, 23.

16. See also G. Feldman, 'Der deutsche Organisierte Kapitalismus während der Kriegs- und Inflationsjahre 1914–1923', in Berding et al., *Organisierter Kapitalismus*, 165; and C. S. Maier, 'Strukturen kapitalistischer Stabilität in den zwanziger Jahren: Errungenschaften und Defekte', ibid. 202.
17. Parker, 'Fuel Supply and Industrial Strength', 156; see also ibid. 157 and n. 4 above (cf. Rathenau).
18. The looser form of company association represented by the *Interessengemeinschaft* was often preferred to complete amalgamation in order to escape higher taxes and to circumvent the regulations regarding the Act on Works Councils which were more exacting the larger the number of employees.
19. *Coalmining: Report of the Technical Advisory Committee*, Cmd. 6610 (London: HMSO, 1945) (also known as the Reid Report); see esp. 18, 32, and 141 for relevant statistics.
20. *Gutachten über die gegenwärtige Lage des Rheinisch-Westfälischen Steinkohlenbergbaus*, Dem Reichswirtschaftsminister erstattet durch Prof. Dr Schmalenbach, Dr Baade, Dr Lufft, Dr Ing. Springorum, Bergassessor Stein (Berlin: Verlag Deutsche Kohlenzeitung, Apr. 1928). The coal industry's official position regarding costs and proceeds is very much open to question (as was, indeed, already evident in the Minority Report presented by the labour nominee, Dr Baade).
21. Ausschuß zur Untersuchung der Erzeugungs- und Absatzbedingungen der deutschen Wirtschaft, *Die deutsche Kohlenwirtschaft: Verhandlungen und Berichte des Unterausschusses für Gewerbe, Industrie, Handel und Handwerk* (III Unterausschuß) (Berlin: E. S. Mittler & Son, 1929); see esp. 167–8.
22. Parker, 'Fuel Supply and Industrial Strength', 157.
23. Ibid. 182. 'Indeed, at times one is tempted to conclude that a fuel industry as such did not exist at all except as a collection of mining interests of industrial concerns' (160).
24. Ibid. 186–8.
25. Ibid. 228.
26. In effect, this 'financial assistance by the banks has greatly accelerated the formation of combines in German industry, while on the other hand the conditions favouring the formation of big combines have attracted and necessitated the assistance of banking finance' (H. Levy, *Industrial Germany: A Study of its Monopoly Organizations and their Control by the State* (London: Frank Cass, 1966; orig. published 1935), 179–80).
27. In 1929, total coal production stood at 124 million tons, 9 million tons higher than the peak pre-war year of 1913; by 1932 production had slumped to 73.5 million tons, the level it had been at the turn of the century.
28. G. Stolper, *The German Economy: 1870 to the Present* (London: Weidenfeld & Nicolson, 1967), 104.
29. Ibid. 115.
30. Ibid.
31. Parker, 'Fuel Supply and Industrial Strength', 188.

32. See also J. Gillingham, *Industry and Politics in the Third Reich* (London: Methuen, 1985), 163.
33. For a fuller account of coal's reorganization see K. Wisotsky, *Der Ruhrbergbau im Dritten Reich* (Düsseldorf: Schwann, 1983), 14 and esp. 36–8.
34. For a thorough account of the relevant legislation and administrative decrees in the 1930s and 1940s, see Boldt, *Staat und Bergbau*, 64–9 ('Die Zeit des Nationalsozialismus').
35. Levy, *Industrial Germany*, 225.
36. See also A. Heinrichsbauer, *Der Ruhrbergbau in Vergangenheit, Gegenwart und Zukunft* (Essen: Glückauf, 1948), 145–6: 'It did not arise under its own initiative but under urgent state encouragement to avoid the inevitable collapse of a few large works with all the evil socio-political and other consequences.'
37. F. Naphtali, *Wirtschaftsdemokratie* (Frankfurt: Europäische Verlagsanstalt, 1966; orig. published Berlin, 1929), 43. See also Levy, *Industrial Germany*, 144, 227.
38. F. Walker, 'Monopolistic Combinations in the German Coal Industry', in *Publications of the American Economic Association*, 3rd Ser. 5: 3 (1904), 322.
39. Levy, *Industrial Germany*, 63, 214.
40. The VIAG (Vereinigte Industrieunternehmungen AG) was a Reich holding company, with a share capital of 180 million marks, responsible for controlling Reich property in banking and industrial undertakings. VEBA (Vereinigte Elektrizitäts- und Bergwerksgesellschaft) was a Prussian holding company, unifying in one organization the financial direction of all the main state concerns in the *montan* and related sectors.
41. Levy, *Industrial Germany*, 226.
42. Naphtali, *Wirtschaftsdemokratie*, 83.
43. Levy, *Industrial Germany*, 223.
44. A more detailed analysis of this phenomenon is undertaken in Part III, Ch. 7.
45. Naphtali, *Wirtschaftsdemokratie*, 51–2.
46. See also ibid. 69.
47. Levy, *Industrial Germany*, 156.
48. Pamphlet of 1926 by the Federation of German Trade Unions, quoted ibid. 211.

PART II

The West German Coal Industry, 1945–1990

3

Coal, Politics, and Industrial Continuity

Part II of this work will indicate that, in the period following the Second World War, the basic pattern of a distinctively German political economy re-emerged in West Germany, almost regardless of the so-called social market economy. This not so new system was characterized by three cardinal elements: a dominant state role; a high level of private industry-wide concertation and co-ordination; and a close co-operation between state and industry. The much-remarked and historically novel phenomenon of a relatively strong and stable trade union movement, fully integrated into the Federal German political economy, will be granted full consideration, but an equal, if not more important, role will be seen to be played by the state. It was primarily state action which was instrumental in saving the West German hard-coal industry from virtual instant oblivion, a potential disaster of such proportions that the whole socio-political system would have been placed in dire jeopardy. System survival was at stake.

The organizational and conceptual basis for the re-emergence of such a system after the war was still very much in evidence in the Ruhr, Germany's industrial heartland. The Ruhr was characterized by a dense interweaving of all economic activity, so well integrated that it displayed an almost organic quality. It possessed an industrial community, to be seen in an interlocking system enhanced by a tradition of collective action which had first been initiated primarily by the private banks back in the mid-nineteenth century, thereafter continued largely by the great cartels, and developed further still after the First World War by the mighty industrial combines, above all, in the *montan*-sector. The scale and importance of the organizational overlapping and common value system of that *montan*-system was to become characteristic of the energy sector after the Second World War, the connecting factor being the coal industry.

But the state tradition was equally, if not more, important. According to Heinrichsbauer,[1] the earlier coal syndicates, for example, had 'only' been emulating the state; where the state intended to plan an organic construction for the economy as a whole, the syndicate had attempted the same within a particular industry. But such a task was only possible on the basis of the pronounced inner cohesion indicated above. The large measure of planning required in heavy industry, not least because of long lead times in investment, sponsored the conscious pursuit by such industry of a form of planning that was a flexible alternative to state planning. This development had by no means been opposed by the state. Indeed, when the great *montan*-combines received compensation payments for plant in the south-west, expropriated after the First World War, they acquired further companies with the full encouragement of the state. Such powerful companies came to be viewed by the state as allies rather than as enemies. They performed an important social function. Thus the creation of the mighty United Steelworks (Vereinigte Stahlwerke) had not arisen spontaneously but as the result of urgent state prompting, to avoid the collapse of certain large companies, with malevolent social and other effects. The state in fact assisted with limited financial support. The parallels with the much later foundation of the Ruhrkohle AG (1969) are inescapable.

Whatever the political and economic context of the German hard-coal industry after the Second World War, its productive prospects were certainly not auspicious. The paucity of coal reserves which were technically recoverable prompted the judgement that 'The Ruhr's strength in coal (therefore) lies in the past'.[2] Since the late 1920s it had been processed coal which kept the industry viable: coke, gas, and chemicals. Moreover, measures were already being taken to spare prime coking coal reserves, mixing the better grades more and more with inferior gas coal. After the war there was also much obsolescent plant and apparently little scope for real improvement by means of rationalization and mechanization, at least at prevailing levels of technology. Nevertheless, a general picture of technological backwardness would be inaccurate; underground mining was as efficient as could be expected, the problem lay with surface workings, especially in electricity generation. But, on the whole, a strong case could be made that the coal industry's prospects were distinctly unfavourable, its demise merely a matter of time.

However, these facts were to be overshadowed by events with international and supranational dimensions which appeared not to have been anticipated in the deliberations of policy-makers. Yet even astute political foresight could not have altered fundamental economic realities. It was highly unlikely that a gradual slimming down of the industry could halt or even reverse adverse cost trends in a world to be increasingly characterized by new and cheaper energy forms, 'Every cut in production (therefore) causes unit costs to rise, whereas they fall when production increases.'[3] This truth was to remain obscured for more than thirty years; there were matters more important than cost structures and the free play of market forces.

Allied attempts in the later 1940s permanently to restructure the German *montan*-industry were to prove to be a total failure. For immensely practical reasons, the Allies had to resort to employing German officials on the ground at a relatively early stage, and this necessarily increased with the passage of time, culminating in the creation of a special agency, the DKBL,[4] entirely staffed by German nationals who were pursuing a most active form of passive resistance. The DKBL embodied the contradiction at the heart of the Allies' intentions for German industry. On the one hand, decartelization was to be pursued seriously, on the other, a German coal organization had been created with a responsibility and power unparalleled in the history of the German *montan*-sector. Admittedly, the DKBL was purely a coal organization, but its personnel were consciously performing a holding operation until circumstances proved more favourable to the re-establishment of customary industrial practices. The Allies had unwittingly taken on a whole industrial culture which, over approximately three-quarters of a century, had acquired very deep roots; its values and coherence proved more tenacious and resilient than the unclear and contradictory plans of the Allies.

The paradoxical dependence of the Allies, and later the ECSC, on German organizations, which were envisaged ostensibly as interim, transitional arrangements but which were to prove both indispensable and enduring, is further demonstrated by the Ruhr coal sales agencies. Although cartels no longer existed, the High Authority not only directly, and later indirectly, controlled coal prices, it also needed an agency on the ground, just as the Allies did, to draw up and execute distribution plans for German coal. The Ruhr coal agencies with their traditions and organizational experience were to prove ideally suited to fulfil such a role.

This organizational continuity was also to be matched by a very early demonstration that the new German state was just as capable of displaying in practice an ideological pragmatism consistent with a long tradition of taking virtually any steps to further national industrial well-being. In January 1952, the CDU-led government, at the insistence of the social market Economics Minister Erhard, introduced its Investment Assistance Act, directed at the whole basic goods/energy sector of the German economy (not just coal). Its purpose was a partial redirection of investment from the manufacturing sector and services into the basic goods industries, situated mainly in North-Rhine Westphalia. In 1950/1 this sector of the economy had been reaching the limits of its productive capacity, representing a severe brake on national economic reconstruction. If investment steering was compatible with the new social market economy, then so was virtually any other kind of publicly directed economic measure.

Although, by the outbreak of the first coal crisis at the end of the 1950s, government involvement in coal industry affairs was considerable, the CDU government was slow to recognize the implications of the sea change manifesting itself on the world energy market. The government's basic attitude was succinctly summed up in a comment by Economics Minister Erhard in the Bundestag: 'We introduced the gas lamp without worrying about the effect on the petroleum lamp.'[5] The advance of oil was not to be hindered out of consideration for the coal industry. The apparently dilatory response of the coal employers was not so much due to nescience or complacency, their action was hampered by ambiguous interest allegiences, not least their intimate involvement in the processing and distribution of oil itself.

The so-called 'zebras', mixed coal and oil companies, were not prepared to put coal first where profitability was concerned. The closing of marginal pits would be made easier where the resources of the oil business could be called upon. But what was most ominous for the mineworker and his family was that it was not necessarily marginal or loss-making pits that would be closed, but very often profitable ones, whether modern or not. Not only that, it would seem that the abortive coal–oil cartel[6] had not been sabotaged by the much-maligned multinational oil companies but by domestic companies.

In fact, the basic situation became extremely ominous for the miners as it was not only mining companies with oil interests that could close apparently sound pits: any new pit, even if owned by, say, a steel company, could be put in jeopardy through falling prices resulting from an energy surplus. Not necessarily the most uneconomic pits would disappear first but those with the severest liquidity problems; given the very high capital intensity and long lead times of the newest pits, they were the ones with the tightest liquidity and therefore the most vulnerable. But for the individual miner, who did not necessarily appreciate the niceties of mining economics, such an argument offered no encouragement. In the period between the two main coal crises in 1958 and 1966, it seemed no pit could be considered safe. Under these circumstances it was not surprising that as the second crisis approached morale in mining communities sank ever lower and that an atmosphere was developing which suggested very adverse political repercussions should tension reach an unbearable pitch.

Combined measures eventually taken to combat the first coal crisis, described in detail in Chapter 5, seemed to achieve some success, with an acceptable balance being struck between the forces for continuity and change. So, for a four-to-five-year period between the two crises, the coal market appeared to stabilize. But under the surface changes were afoot with adverse consequences for miners, the most important of these being, first, their loss of the premier position at the top of the league table of industrial workers' wages and, secondly, the increase in accidents as the pace of mechanization was forced in the drive for increased productivity, sometimes with machines not having been adequately tested in advance. However, these developments were not so immediately apparent at the time. More noted perhaps were the freeing of miners with two or more children from having to pay income tax, and, from January 1961, the release from national service obligations of underground workers, developments in the long line of privilege which has set the miners apart from other German workers.

Yet despite the apparent stabilization of the industry, miners voted with their feet. Throughout this period there was a shortage of skilled workers and this was also the time when foreign workers came to be employed in the industry in substantial numbers.[7] A rapid increase in their employment began in 1961 at those more marginal pits which could not offer to pay the premium rates offered by the more viable ones. By 1963, more than two in three new miners were foreigners, a trend not unrelated to the deteriorating earnings of miners overall and to the increased accident rate due, for instance, to language difficulties in operating complex new machinery. The Federal Labour Office co-operated with the coal employers in the hiring of foreign workers by often making the work permit valid only if they initially agreed to work and then stay in coalmining. The *Gastarbeiter* phenomenon in mining was to remain a permanent feature of the coal industry's further evolution. In the wider economy, however, a more significant development was underway: the drive towards forms of co-operation which were, nevertheless, supposed to be compatible with the competitive doctrines of a market economy in general and consistent with the provisions of the Act on Restraint of Competition in particular. This development represents as much of a turning-point in the Federal Republic's history as the much more widely commented reassertion of a dominant state role in the management of the economy as embodied in such measures as Schiller's Law for the Promotion of Stability and Growth in January 1967.

The re-emergence of that powerful feature of German capitalism, close collaboration rather than sharp competition, revealed the strength of a tradition which was too strong for government to resist. Indeed, on the contrary, the state appeared actively to sponsor industrial collusion. This was to be demonstrated by a major initiative conducted by the government of West Germany's premier industrial state, North-Rhine Westphalia. The approach was enshrined in two related policy documents in the mid-1960s,[8] essentially recommending more of the same, the same being increased collaboration, particularly between government and industry. Inter-company co-operation was also to be given urgent stimulation, concerning both vertical and horizontal linkages; this was to apply particularly to small and medium-sized companies (*Mittelstand*). Special weight was granted to those various areas of co-operation not covered by the restrictions of anti-cartel law. The individualistic prejudices of entrepreneurs were to be combatted and publicity campaigns were to be inaugurated to engage the support of the wider public for co-operative values.

Greater industrial co-operation at this time was also reinforced by the collaboration, indeed exceptional harmony, between the employers and the unions. It found specific expression in its endorsement by the prominent right-wing social democrat and senior trade unionist, Georg Leber, MdB. The old opposition of capital and labour was to be resolved on the higher level of a property-owning democracy,[9] through a natural process of evolution towards unity. Apparently, this vision did not require parity co-determination as a necessary, let alone sufficient, condition for its fulfilment. Yet, in the view of the coal employers, *montan* co-determination did fulfil an important informational function,[10] and works co-determination in particular was to represent not only a stabilizing function but also a conflict-reducing, harmony-promoting institution. This will be seen to have been the case during the first coal crisis and its aftermath. Co-determination was so significant here not least because the employee side of it was tightly controlled by IG Bergbau, exercising an order maintenance role as was expected by the employers.

It will also be demonstrated that, by the time the second coal crisis broke, a clear triangle of basic interests had emerged: the German iron and steel industry, the state, and coal, the coal interest being expressed primarily by mining labour, in practice by the IG Bergbau. The coal employers could not represent a specific interest by themselves. Their involvement in the oil business and the power structure of coal ownership in which the iron and steel interest dominated precluded the formation of a forceful, autonomous coal employer interest. On the other hand, the mining union was both exceptionally well organized and unhindered by other commitments. Its purpose was single-minded, to prevent job losses, although not at any cost, and to maintain and improve the welfare of miners. It received political backing from the SPD and was invariably subject to very sympathetic consideration and support by the general public. Nor was the other major political party, the CDU, hostile or indifferent to the plight of the mineworkers; too much was at stake electorally, with so many votes involved, both at *Land* and Federal level.[11]

But the key to the union's power was to lie in its pivotal position in the whole political economy of the Ruhr. This once large and still moderate union was playing a very constructive role both within the organs of parity co-determination and as an industrial union, effectively representing miners irrespective of their religious beliefs and political commitment. For government to eschew this union and its concerns would be to spurn the positive contribution of organized labour to the economic and political regeneration of free Germany and would perhaps guarantee a widespread socio-political alienation which was, in fact, to begin to emerge as the recession developed in the mid-1960s. Unions were conceived as a force for order; one way to guarantee social peace and political stability was to win and retain the loyalty of organized miners. Permitting the union genuine exercise of influence in the relevant councils and counsels was a reasonable price to pay in order to maintain the high level of industrial, social, and political integration obtaining in the Ruhr, the foundation of system survival.

But none of these three interests exhibited an automatic devotion to conventional market solutions. It was inconceivable that the union would demand the rigorous application of market rules and although the state had now become the primary guardian of the competitive order, its responsibilities were wider and more comprehensive than maintaining complete competition (*vollständiger Wettbewerb*) in a particular sector of the economy. Experience in the Federal Republic was to demonstrate that the state tradition, of appearing outside and above the economy (despite the plethora of practical connections) had not been materially affected. A strict adherence to market doctrines would represent an ideological inflexibility incompatible with the state's function as honest broker in a highly complex situation. An harmonious resolution of conflict was to be attempted on the basis of reconciliation of interests, for the benefit of the whole.

The domestic survival of the basic iron and steel industry was also in question at this time, at least in its prevailing form. This was an industry which neither displayed nor demanded the textbook application of a market philosophy, such would have been far too incongruous, given the realities of that industry's organization, mode of operation, and certain technological trends at work.[12] In the event, the iron and steel industry was to extract a very good deal from government. First, it benefited from the (ECSC) coking coal subsidy, the basis of the German coal industry's ability to supply West European heavy industry with coke. Secondly, the closure grants, premiums, paid also to pits owned by steel companies, meant that they had that liability recompensed. Thirdly, and above all, the capital released by the sale of mining property, mainly land to the Action Community,[13] was to represent a very welcome source of funds for vitally required investment in an area of intensified international competition.[14] Furthermore, the structure of the Ruhrkohle AG gave the steel companies a significant place on the Supervisory Board, an influence consolidated by the so-called Foundry Contracts.

The state rather than the market clearly emerges as the fulcrum of developments in the coal industry. Co-operation between the state and the steel industry was similarly close, indeed, one insidious example was the connivance of the Federal government and the *Land* government of North-Rhine Westphalia with the *montan*-industry in the early 1960s, preventing the location of new industry in the Ruhr. It was feared that new jobs would be likely to lure workers away from the basic industries and such collusion precisely contradicted efforts of the ECSC's High Authority to initiate employment diversification in coal areas. It would seem that all roads lead to the state in the West German political economy. Not only was the state called upon generally to prevent structural crises, the present study clearly demonstrates how in the coal industry it attempted to reconcile the varied interests, establishing a common denominator as the basis for some kind of plan, in the interest of a wider whole.

The state even interfered in the sphere of wage negotiations,[15] so significantly, that it would seem that, in the coal industry at least, the much-vaunted tarif autonomy was a very relative concept. In general terms, state involvement usually entailed picking up a bill of some kind, especially in the field of investment.[16] By 1974, Federal expenditure to assist the coal industry was to rise to more than 1.5 billion DM, compared with

909.8 million DM in 1973. Over the twelve-year period, 1964–76, public support for the coal industry was to amount to nearly 17.5 billion DM: 8.139 billion in various direct grants (*Finanzhilfen*) and 7.050 billion in tax concessions.[17] This total figure includes neither *Länder* and local authority support nor support through non-nuclear research, regional, and town and country planning (*Raumordnungspolitik*) programmes. It would seem that the state had clearly become the coal industry's paymaster, but, as will be demonstrated, no simple, straightforward channel for exerting state influence appeared to exist, especially as nationalization was not a realistic option.

Indeed, formal mechanisms of public control proved to make very little impact, whether in coal or elsewhere in the energy sector. This applied to gas supply and investment, despite powers granted to *Land* authorities by the 1935 Energy Act; and local authority involvement seemed also to make very little difference in the sphere of electricity production and distribution. This phenomenon is probably associated with the fact that most energy companies in the Federal Republic are multi-product suppliers (*Mehrproduktenanbieter*). Nowhere is this more emphatically illustrated than by VEBA, at that time the largest single shareholder of the Ruhrkohle AG, which merged with Gelsenberg in 1975. In fact, although VEBA's main area of activity was energy supply, it also possessed major interests in chemicals, glass, commerce, and transport. As the state (*Bund*) held a 44 per cent share in VEBA, the state (broadly interpreted as 'die öffentliche Hand') controlled by means of its shareholdings 51.6 per cent of German hard-coal production. Including the Saar, this public influence accounted for 85 per cent of total production.[18]

However, such formal considerations, important though they are, do not present as clear a picture as might be supposed: 'the Federal Government and the *Länder* do not use their industrial holdings as a state energy policy instrument.'[19] More important, though, is the fact that 'in essential aspects . . . the energy sector consists of large companies which, directly or indirectly, are so enmeshed with one another through capital links of subsidiary companies that it is practically impossible in many cases to identify the actual centres of decision'.[20]

This difficulty in locating the precise centres of power necessarily renders it extremely difficult to allocate and evaluate clearly the respective weight of public and private influence. But that the scale and nature of that influence was to be of vital importance is evident from two facts. First, the percentage of the energy sector's turnover of total national turnover was growing fast, reaching 10 per cent in 1974 (and 15 per cent of total industrial turnover in 1973). Secondly, the energy sector was critical, for the remainder of the economy depended on it: growth, full employment and consumption.[21] Moreover, 'associated companies [*Gemeinschaftsunternehmen*] in the energy sector constitute an important form of industrial planning',[22] but the concentration of decision-making in a few hands was diametrically opposed to a main thrust of social market doctrines, the decentralization of economic power throughout society. Traditionally, the coal industry had embodied associated companies in their purest form, and the creation of Ruhrkohle AG was to appear to represent the culmination of the concentration of coal interests.

The exercise of public power in energy generally, but particularly in coal, includes but transcends questions of ownership. Shareholdings are important but are not the real basis of state influence; the latter flows from its role as banker of last resort, especially in regard to coal, and in the 1970s and 1980s this was to be as apparent as ever. In the absence of nationalization and of active use of state shareholdings the fiction of the disciplines of a market economy could still prevail while the German political economy continued along its customary path, one of mutual accommodation and one that eschewed brinkmanship as a principle of negotiation, the sense of mutual dependence being too great.

The state role was to prove a conciliatory one, for even when the state possessed coercive reserve powers, it preferred not to employ them assertively. The compulsory purchase powers of the Coal Adaptation Act, designed to back up the work of community action, were employed with restraint, with very few cases disputed. The controversial power of the Coal Commissioner was to be exercised with consideration, and his much-questioned power of financial sanctions to enforce recommendations was seldom employed in practice. The provisions of the Coal Act also permitted the Coal Commissioner to accede to trade union demands for slowing down coal's decline in the interest of minimizing redundancies; the sensitive issue of pit closures was not to be resolved unilaterally but on the basis of concertation.

The active desire to promote agreement by resorting to less formal means of concertation were by no means to be restricted to the coal industry; it would be characteristic of relations with the energy sector as a whole and was to become ever more refined as the 1970s progressed. Haupt *et al.* sum up the position clearly:

Co-ordination between the state and industry in the energy field is largely in non-institutionalised form between individual companies or associations and the

Federal Ministry of Economics. These discussions and negotiations—in contrast to those in some other economic sectors—are not formally linked to advisory councils or other bodies. But with the growing role of the state in energy policy the informal contacts will certainly have increased both in number and political significance since 1973 and their weight should not be underestimated.[23]

Their purpose would be to promote private sector agreements which would make prospective government measures superfluous.

Government relations with the public sector were hardly different from those with the private sector. Public enterprises in the energy sector, and particularly VEBA, operated on private lines, and, in practice, local authorities did not differ greatly from private owners in the energy supply field. Again Haupt *et al.* summarizes the position as regards political concertation:

Co-ordination between the various public institutions, especially between the Federal Government and the Länder, is within the framework of the regular contacts which have existed for some time, whether on the level of the specialised personnel or on the political level. These forms of co-ordination are in general use, they are not limited to the energy sector.[24]

So the well-established and largely informal channels characterizing relations within the public sector, between the public sector and government, and between the state and private industry were not exclusively typical of the energy sector: the employment of the methods of hearings and non-institutionalized informal contact with the private sector were in use before the oil crisis and they were also usual outside the energy sector.

Although there were to be no institutional changes in decision-making procedures regarding energy policy as a result of the energy crisis, either at Federal, *Land*, or local authority level, an intensification of an intermittent yet apparently inexorable growth in state responsibility and activity in both *montan* and energy spheres can be clearly demonstrated, at least until the end of the 1970s. This was to be especially apparent regarding investment, indeed, there was a form of investment-steering which was equally apparent in other industrialized countries. Yet this growing state involvement was to find little formal institutional expression; it primarily takes the form of an intensification of existing informal contacts with interested parties:

the higher political importance which decisions 'on the energy questions' had acquired was apparent in that higher and top level political bodies increasingly concerned themselves with decisions which were generally taken on lower levels.[25]

Energy developments were setting precedents for other areas.

Although many formal adherents of the social market economy were naturally reluctant to concede a major and growing role in the economy for the state, the evidence to the contrary was to become overwhelming and conclusive. The institution of the state has shown a remarkable tenacity and endurance, demonstrating an unmistakable continuity with the German tradition of political economy. For various reasons, state influence was to be exercised largely through informal channels, and the fact that they seemed on the whole to operate successfully is a testimony to the conciliatory response of capital and labour, evidencing a readiness to co-operate which may be taken to testify to a common value system which transcends such elemental conflicts as social class. Perhaps this phenomenon expresses an awareness of a common interest, namely Germany's success as an industrial nation. There was to be little doubt that the primacy of trading relationships dominated the thinking of Federal German policy-makers,[26] and not just in exclusively economic spheres— it is this enduring priority which provides the basis of the state's virtually unbroken interest in, and active concern for, economic welfare.[27]

This concern, of a kind apparent for a hundred years, was as valid as at any time in Germany's history, if not more so. The rise of Japan, for example, was a sign of what was perhaps to come in the way of challenge from new competition from the Far East. In short, 'Energy prices are ... of decisive importance for the price level of a national economy and its international competitiveness'.[28] And with the broader energy crisis, energy policy had also become inextricably entwined with considerations of balance of payments equilibrium thanks to West Germany's high energy consumption and limited natural resources.

Industrial concentration, in the interests of combining resources to promote higher productivity and growth, was very much at odds with the main decentralizing thrust of the social market economy, but it was required in the interests of international competitiveness. In point of fact, the necessity to co-operate seems to be the natural reaction in solutions to Germany's economic problems. Market doctrines or not, competition, at least domestically, is divisive, representing dissipation of strength. Co-operation implies unity, the gathering of powers to fulfil the common purpose, namely success in foreign markets which is the key to stability, growth, and well-being at home. Hence the need for a prominent role for the state, as the arch interest-reconciler, the state is the catalyst for that concertation and co-ordination which appears essential for national success—market doctrines, in a narrow sense, must take second place. Given the prevailing ideology of the social market, the centripetal, integrative force of the state is required to balance the centrifugal, atomizing influence of competition. But an opposition of state v. market/competition should not be overstated, it is largely circumvented in practice by state–private industry collaboration. The re-emergence of Germany's collaborative capitalism in the 1960s was to be actively supported by public bodies, and to this end public financial support was to be made available.

Whatever the precise ramifications and implications of publicly promoted industrial modernization and domestic co-operation in industry, West Germany's belated and deficient anti-monopoly legislation was not allowed to represent an obstacle to close collaboration. By the 1970s, at the latest, it would seem that, on any occasion where cartel legislation appeared to threaten the degree of co-operation and co-ordination which was deemed necessary in the context of international viability, no compunction was to be shown in demanding and effecting its modification.[29] The energy sector had acquired the strategic significance that the *montan*-sector possessed before the war; its requirements were simply not to be handicapped, the imperatives of international success were paramount.

But where did all this leave the coal industry? A cynical observer might comment that during the 1970s it was to degenerate into merely a coal-stocks holder for iron and steel and for the electricity industry which was demonstrating its dedication to German coal's interests by undertaking the pursuit of pit acquisition in North America and Australia. At least clarity would obtain in one respect, it had to be officially admitted that the German coal industry owed its existence to state support: 'Coal mining . . . had been dependent on state aid since the end of the 1950s.'[30]

So the fate of coal would be decided by the state, but would the state assess coal's place according to its possible contribution to national economic well-being? Apparently not. The case of coal is an illustration of the triumph of politics over economics, or, more precisely, the indissoluble nature of them both; as long as the coal industry poses a sociopolitical threat to German democracy it will be preserved. In fact, a surgical operation against coal is inconceivable for other reasons. Just as the West German economy has been characterized by its interlocking, organic quality so the mining industry has been wholly embedded in an equally complex and intricate web of political links that have sustained it in such a way as to make its run-down seem a destructive act of (social) self-mutilation.

Notes

1. See A. Heinrichsbauer, *Der Ruhrbergbau in Vergangenheit, Gegenwart und Zukunft* (Essen: Kettwig, 1948), 97.
2. Ibid. 29.
3. G. von Beckerath, 'Kostenstruktur und Marktordnung im Steinkohlenbergbau', *Zeitschrift für handelswirtschaftliche Forschung*, 8 (1956), 242.
4. Deutsche Kohlenbergbau-Leitung.
5. Quoted in W. Abelshauser, 'Von der Kohlenkrise zur Gründung der Ruhrkohle', in U. Borsdorf and H. Mommsen (eds.), *Glück auf, Kameraden! Die Bergarbeiter und ihre Organisationen in Deutschland* (Cologne: Bund-Verlag, 1979), 427.
6. See p. 99.
7. Their number rose steadily from 9,458 in 1959 to 23,694 by Mar. 1964—this represented an average of the total work-force of 4.3%; but these foreign workers were only employed underground, so the average underground became 9% by the end of 1963 (in some pits the percentage was to rise to 34%).
8. Minister für Wirtschaft, Mittelstand und Verkehr des Landes NRW (ed.), *Notwendige Maßnahmen zur Verbesserung der Landesstruktur in NRW*, i. *Analyse und Vorschläge zur regionalen Strukturverbesserung* (Düsseldorf: 1964). and Minister für Wirtschaft, Mittelstand und Verkehr des Landes NRW (ed.), *Notwendige Maßnahmen zur Verbesserung der Landesstruktur in NRW*, ii. *Strukturveränderungen durch neue politische, wirtschaftliche und technische Entwicklungen* (Düsseldorf: 1966).
9. See G. Schneider, 'Neue Allianzen für die freiheitliche Ordnung', in *Zwang und Grenzen der Konzentration* (Düsseldorf: Handelsblatt, special edn., Aug. 1966), 59.
10. The satisfactory conduct of wage negotiations in 1963 had been possible because union negotiators had been fully informed about the financial situation of mining by their representatives on management and supervisory boards.
11. This is not to imply that Christian Democrat concern was purely opportunistic: on the contrary, there had long been an intimate Centre party involvement in mining affairs to which the CDU was heir, and Heinrich Imbusch's Christian mining union had been an active and forceful labour union in the pre-war years.
12. One such change related to the importance of iron ore in the basic production process growing in relation to that of coal.
13. See p. 111.
14. In a mere five years, 1960–5, the Japanese share of the EEC steel-market rose from 7.5% to 20.4%.
15. See C. D. Schmidt, 'Die Krise im Steinkohlenbergbau und ihre soziale Problematik unter besonderer Berücksichtigung des Ruhrgebietes', Ph.D. thesis, Westfälische Wilhelms-Universität, Münster, 1967, 187 and 195.

16. V. Bahl, *Staatliche Politik am Beispiel der Kohle* (Frankfurt: Campus-Verlag, 1977), 59.
17. W. Mönig *et al.*, *Konzentration und Wettbewerb in der Energiewirtschaft* (Munich: R. Oldenbourg Verlag, 1977), 867.
18. The CDU government's privatization of VEBA in the 1980s was not to represent a major departure in either energy or coal policy.
19. R. Haupt, H. Mirbach, and H. Siedentopf, *The Adjustment of Administration to the Energy Crisis*, International Symposium Report, Brussels (Bonn: Federal Ministry of Economics, 9–11 June 1982), 28.
20. Mönig *et al.*, *Konzentration und Wettbewerb*, 52.
21. U. Dolinski and H. J. Ziesing, *Sicherheits-, Preis- und Umweltaspekte der Energieversorgung* (Berlin: Duncker & Humblot, 1976), 22.
22. Mönig *et al.*, *Konzentration und Wettbewerb*, 15.
23. Haupt *et al.*, *Adjustment of Administration*, 11.
24. Ibid. 29.
25. Ibid. 27.
26. Cf. C. Deubner, 'Change and Internalization in Industry: Toward a Sectoral Interpretation of West German Politics', *International Organization*, 38 (1984), 501–35.
27. Economic Advisory Council to the Economics Ministry, *Wirtschaftliche Folgerungen aus der Ölverknappung* (Bonn: Federal Press Office, Bulletin No. 155, 20 Dec. 1979), 1423.
28. Dolinski and Ziesing, *Sicherheits-, Preis- und Umweltaspekte*, 23.
29. See e.g. J. Junghans, 'Zur leitungsgebundenen Energiewirtschaft', in H. Anders, K. H. Hoffmann, and F. Voigt (eds.), *Energie: Leistungen Prognosen Alternativen* (Mannheim: Mannheimer Verlagsanstalt/Verlagsanstalt Courier, 1972), 40.
30. Haupt *et al.*, *Adjustment of Administration*, 21, continuing: 'which was very largely determined by employment and regional and structural policy considerations' (28), i.e. energy policy in general and security of supply in particular played little or no role in the decision to protect coal.

4

The Rise Before the Fall, 1945–1957

1945–1948: Dislocation and Improvisation

In April 1945 British and American troops entered the Ruhr, and on 8 May 1945 German troops unconditionally surrendered to the Allied Powers. In a joint declaration of 5 June the four Allies assumed ultimate state authority in Germany. In the Western zones two problems weighed heavier than the actual level of wartime destruction of industrial capacity: the totally disrupted and disorganized transport network and the need to feed, clothe, and shelter not only the indigenous population, but also the huge number of refugees and expellees from the Eastern zones. The mining industry's housing stock had suffered particularly badly. Of 311,362 dwellings, 237,492 had either been destroyed or damaged. Underground workings had been largely spared, but above ground installations had been severely affected, especially cokeries, by-products works, and, above all, colliery companies' power plant. Although the number of employees had remained substantially unaltered: 297,000 in 1936 (FRG area),[1] and 276,192 (Ruhr), in 1945 their productivity had fallen substantially for two reasons. On the one hand, the high average age of miners due to younger miners being killed in the forces and the lack of youths in training as their successors; on the other, approximately two-fifths of the industry's labour force at the close of the war consisted of foreigners, forced labour from occupied territories, and prisoners of war who were understandably not as industrious as native workers. For the whole of 1945, Ruhr coal production amounted to only 33.1 million tons and coke production was down to 5.1 million tons.

One of the Allies' first steps regarding the coal industry in the Western zones was to seize all mines and take over their administration on the basis of Law No. 52 of the Military Government. Overall control was provided by the Solid Fuel Section of the Supreme Headquarters of the Allied Expeditionary Forces; the subsection responsible for the Ruhr was known as the Rhine Coal Control and was divided into six regional offices. Once final clarity had been established on the precise demarcation of the occupation zones the British military government established as the supreme coal body the North German Coal Control (as the successor to the Rhine Coal Control), with its seat at the Villa Hügel in Essen. One of its first moves was to dissolve the Rhenish-Westphalian Coal Syndicate and the Rhenish Lignite Syndicate on 6 September 1945. The transactions of the former organization were transferred to the Ruhr Coal Distribution Office (RCDO) and those of the latter were transferred to the Coal Distribution Office in Cologne.

On the basis of the Law No. 52 the British Military Government issued General Order No. 5 on 22 December 1945. It concerned the blocking and control of assets in hard-coal and lignite mining, making all collieries and associated works in the British zone liable to 'confiscation, direction, administration, supervision or other control'.[2] Colliery owners ceased to exercise any rights, for example meetings of the supervisory board and of the AGM were rendered impracticable. Although property relations were not regulated it was initially stated that German owners would never be allowed to resume possession of their property;[3] however, precisely one year later, controls were considerably relaxed.

The Ruhr Area Group, which had been established during the Third Reich, had to be disbanded. The employers' technical association, the Mining Association, was suspended and its business assumed by a trustee. Nevertheless, the practical effect of these steps was more apparent then real. On 8 December 1946, the first miners' trade union was formed in Bochum: the Mining Industry Association (Industrieverband Bergbau), changing its name to the Industrial Mining Union (Industriegewerkschaft Bergbau/IG Bergbau) in 1948. The Association of Senior Mining Officials (Verband oberer Bergbeamten), which had been disbanded in 1933, was reconstituted on 24 April 1947. By the end of 1946, although yearly coal production had reached 50.5 million tons, it was still less than that of 1900.

The North German Coal Control was headed by British officers and production matters were administered by German mining engineers, accountants, and miscellaneous officials drawn from the former Mining Association. Yet despite the earlier arrest of top Ruhr industrialists and the apparent dissolution of all joint industry-wide organizations in German coal-mining steps were being taken to ensure continuity:

before the creation of the DKBL,[4] the resolute delaying tactics and shrewd operations of leading personalities within the circles of the previous communal associations [*Gemeinschaftsorganisationen*] succeeded in keeping together the basic organisation and in preventing Allied actions from making coal mining a vacuum to be filled by experiments and unrealistic 'constructions' dictated by Allied interests.[5]

The German mining experts knew how to exploit the Allied dependence on their expertise and time was of the essence in their determination not to have a fundamentally new form of industrial organization imposed upon them. This resolve was strengthened by the agreement between the British and Americans to unite their zones economically, and the bi-zone came into effect on 1 January 1947.

Nearly two years after the end of the war the general food situation had deteriorated even further and in April 1947, after a particularly severe winter, the Ruhr coal miners went on hunger strike.[6] At least, coal had been released for household consumption during that winter, for the first time since the war. Their allocation of ration points for vital necessities, already better than for workers in other trades, was consequently increased. In particular, extra stamps for groceries were awarded and care-parcel campaigns were stepped up for their benefit. When the miners returned to work they had achieved more than an improvement in the level of their own provision, they had forced the Allies to become explicitly aware of what had been implicit for some time, that the German economy could not function properly without a sound and thriving coal industry. Perhaps, more importantly, it was also recognized that a functioning German *montan*-sector was indispensable for the reconstruction of the whole of the Western European economy. This realization coincided with the final breakdown of inter-Allied negotiations on resolving the general question of reparations and the particular matter of control of the Ruhr (Foreign Ministers Conference in Moscow, 16 March–24 April 1947). Within two months the Marshall Plan was announced and a mere month after that the Organization for European Economic Co-operation was created to implement the European Recovery Programme.

The Allied decision, under American initiative, to promote a programme of reconstruction on a consistent and sustained basis required the co-operation of indigenous authorities and personnel. The newly created Economic Council (*Wirtschaftsrat*), under the chairmanship of Professor Ludwig Erhard, was empowered on 18 November 1947 to pass laws and issue directives in certain coal-related spheres.[7] Article One of the relevant Decree established an organization consisting entirely of Germans, to direct all parts of the coal sector: the Deutsche Kohlenbergbau-Leitung (DKBL), based in Essen. This organization was responsible both for hard-coal and lignite mining. It was generally responsible for the efficiency of the industry, in particular for directing the production, shipment, and dispatch of coal, coke, and all other by-products. The head of the organization was managing director, Dr Heinrich Kost,[8] who was appointed by the Allied Military Government. He, in turn, appointed his departmental heads in consultation with the Allies— all other appointments lay entirely within his own discretion. As the DKBL consolidated its position, the old sense of joint responsibility reasserted itself in the newly constituted expert committees and working parties.

During the same month, November 1947, the North German Coal Control was joined by American representatives and the organization was renamed the United Kingdom/United States of America Coal Control Group (UK/USCCG). It was later joined by French representatives on 1 March 1949 and then simply known as the Combined Coal Control Group. At the same time the North German Coal Distribution Office was incorporated into the DKBL, becoming renamed the German Coal Sales (Deutscher Kohlen-Verkauf), the central selling agency of the DKBL. The authority of the DKBL over mining companies was to become total: it had become a kind of super-cartel, with a power unique in the history of German mining industry organizations.

In June 1948 the Currency Reform banished the danger of the kind of hyper-inflation which had had such fateful political consequences in the Weimar Republic. But coal was to remain an exception: its prices were not de-controlled by the Law Regarding Guidelines for Industrial and Price Policy after the Currency Reform of 24 June 1948, the reason being that coal, like other raw materials and basic foodstuffs, formed 'an essential basis for the industrial and agricultural production of goods'.[9] Between 1932 and 1948 prices had been pegged at roughly the same level: between 1 May 1941 and 1 April 1948 there had been no increase at all. Two price rises were permitted in 1948, together increasing coal prices by 214 per cent, nevertheless, according to the report of a commission of inquiry for 1948, 71 per cent of all Ruhr coal-mines were running at a loss. Export prices of German coal were to remain under complete Allied control.

General price de-control (associated with the Currency Reform), the native supervision of controlled exceptions, the unmistakable re-emergence of determined and competent indigenous organizations to manage the coal industry, and the establishment of appropriate labour organizations, all combined to ensure that German interests would make themselves clearly felt from then on. Western European dependence on German coal consolidated the influence of these domestic interests.

1948–1952: New Foundations

The Allied recognition of the necessity to involve indigenous West German authorities in the reconstruction of the economy, both to relieve the administrative burden on the Allies and to permit a positive contribution towards the reconstruction of the Western European economy, became an irrevocable decision with the occurrence of the Berlin Blockade and Airlift. This event symbolized the final breakdown in East–West relations. Nevertheless, the Allied resolve to break the power of the mighty *montan*-combines of the Ruhr was not materially affected. But the determination of the Allies to destroy, or at least effectively control, what had been the basis of the German war machine in two World Wars was not matched by clarity on how to achieve this vaguely defined objective. There was no consensus among the Western Allies on precisely how to achieve specific objectives; conflicting interests were in a constant state of flux. Decartelization took pride of place. In 1945 there had been general Allied agreement on three definite priorities: severing the links between iron and coal; generally reducing the size of coal and iron and steel companies; and substantially reducing, if not eliminating, their influence on coal by-products works and on iron- and steel-processing companies. In December 1945 the links between coal and iron and steel had been severed—Operation Severence. Separate Coal and Steel Control Groups were set up, they were kept entirely separate and for all practical purposes were considered independent of one another. It was decided at an early date that the regulation of property relations was to be left to the decision of a sovereign elected assembly of Germans.[10]

An attempt was made to remedy anomalies with Law No. 75 of 10 November 1948. It represented the first clear attempt to define actual measures of decartelization and organization with an appropriate legal basis. The Law pursued a dual purpose: first, to prevent the re-establishment of the kind of ownership relations which would represent an overlarge concentration of economic power; and, secondly, to restructure coal-mining in such a way as to further the development of the West German economy as a whole. This was the first time that largely economic considerations seemed to motivate the design of such industrial regulations; the construction of production units of optimal efficiency was to be the decisive consideration in the practical execution of the new regulations. Nevertheless, all decisions and measures remained subject to sanction by the Allies.

However, two immensely practical and apparently intractable problems emerged. First, the West German authorities were still implacably opposed to severing the organic links of the established *montan*-combines, and the longer they employed delaying tactics the more likely was the possibility of outlasting the Allies. Secondly, precisely who would purchase these new coal companies (an even more problematic issue for the new steel companies)? The Allies were confronted with the desolation of the tenuous capital market, the great uncertainty still surrounding the future ownership rights and compensation entitlement regarding the stocks held in trusteeship, and the uncooperative hostility of the West German coal authorities to industrial outsiders.

In the meantime, the allocation of total German coal and steel production between exports and domestic consumption had become a subject of renewed international attention. Consequently, the International Authority of the Ruhr (IAR) was established in December 1948, on the basis of the Ruhr Statute. It represented a form of internationalization of the Ruhr whereby the United States, Britain, France, and the Benelux countries jointly undertook allocation. West Germany was awarded three seats on this twelve-man body and this was the first post-war example of official German representation in an international organization. The primary purpose of the Statute was to ensure the equal and unhindered access of West Germany's neighbours to its coal, iron, and steel and this was specifically intended to pre-empt the possibility of renewed German aggression and to prevent Germany from being able directly to determine the pace of Western European reconstruction. Supplies of Ruhr coal in particular were guaranteed to the IAR members at a time of general coal shortage and apparently at prices well below the prevailing world level.[11]

Increasingly, West German nationals were also given greater control over coal distribution. The process of handing over the actual allocation and distribution of coal for German consumption to German agents had begun as early as 1 January 1947 in the British zone, and even earlier in the American zone. Under Allied supervision distribution plans were drawn up by the Economic Administration (*Wirtschaftsverwaltung*, a wholly German body) and details of implementation were the responsibility of the Land Economics Ministries, working in co-operation with the marketing organizations. The system of rationing was liberalized after the Currency Reform; from the autumn of 1949 controls were relaxed even further and all restrictions were removed for industrial consumption in January 1950 and for household purchasers by the end of March.[12]

By 1950, there was a broadly satisfactory rise in coal production up to 110 million tons for the area of the FRG, although this was still well below the 136.9 million tons of 1938. But the Allies were still substantially dissatisfied with the lack of effective progress in fundamentally reorganizing the *montan*-industry, through genuine decartelization. Law No. 75 had not inaugurated the hoped-for improvement in an industrial restructuring consonant with the determination of the Allies to deal once and for all with the concentration of heavy industry in the Ruhr. A final Allied attempt found expression in the long, comprehensive, and highly detailed Law No. 27 of the High Commission Regarding the Re-shaping of German Coal Mining and German Iron and Steel Industry, promulgated on 16 May 1950. The Allies now appeared to be determined to pursue decartelization as a means of decentralizing the economy which in turn was intended to stimulate economic activity and growth in a way compatible with, and conducive to, political democracy. In other words, Allied policy was committed to furthering the positive aims of politico-economic development by means of decentralization and decartelization, the latter clearly no longer a means of restricting and restraining industrial potential but rather its opposite.

On 15 September, with the blessing of the Federal government, the DKBL dutifully presented a pure coal plan. It envisaged twenty-three colliery companies, with an output ceiling of 9.4 per cent of total national output for any one company. Yet this plan, like others before it, was to remain a plan. Indeed, the psychological war of attrition of the agents of the Ruhr industrialists against the Allies in regard to preventing the complete disassociation of coal and iron and steel finally began to bear fruit in 1951. The Allies did not automatically dismiss the joint plan, the *Gemeinschaftsplan* of the DKBL and the Steel Trustees which was submitted in January 1951. It envisaged the iron and steel industry possessing an overall 24 per cent stake in coal-mining companies accounting for 30 per cent of West Germany's coal production. The Allies submitted a counter-proposal which insisted that no single steel company could obtain more than 75 per cent of its fuel requirements from pits which it owned itself. The Federal government, which had been directly involved in the negotiations since its inception, was obliged to play a mediating role and accordingly submitted compromise proposals on 14 March 1951. With the necessity to compromise apparent to both Allied and Federal German governments, the Allied High Commission accepted these proposals on 27 March 1951, endorsing the 75 per cent principle, with certain modifications.

Probably the most important factor promoting a meaningful *rapprochement* among competing national and international interests towards the end of 1950 and into 1951 was the existence of the Schumann Plan.

The Plan, eventually forming the basis of the European Coal and Steel Community, appealed to the Federal German government partly because it offered continuing support for at least some kind of anti-trust policy, but primarily because it represented a unique opportunity for the acceptance of the Federal Republic into the international community of nations on a full and equal basis. The interests of German's Western neighbours were both political and economic. On the political side was the need to maintain some form of adequate international control over Germany's *montan*-complex, in the vital interests of their own security. Economics emphasized the need to obtain guaranteed access at the lowest possible price to Ruhr coal upon which their own economies to a large extent depended. Realism, in the form of guarding essential national interests, combined with idealism, the spirit of European co-operation, and created an irresistible force making for mutually advantageous agreement. The treaty founding the European Coal and Steel Community, signed in 1951, presupposed equality of treatment of all partners. To this end the West German authorities greatly welcomed the opportunity to throw off the limitations of the Ruhr Statute and the IAR, both of which duly disappeared when the ECSC Treaty came into effect on 25 July 1952.

In 1951, 444,934 Ruhr miners had produced 110,630,000 tons of coal, with an output per employee of 249 tons; by 1952, 462,715 miners were producing 114,417,000, with two more pits (making 142), and an individual output of 247 tons. Indeed, 1951 had proved to be a very eventful year.[13] In May 1951 the Act Regarding Employees' Co-determination on the Supervisory Boards and Boards of Management in the Iron and Steel Industry and in Coal Mining became law. This institutionalization of parity co-determination had initially been established in the German steel industry in 1947, and its extension to the coal industry occurred as the result of a determined and sometimes bitter struggle by the labour movement to retain this major advance for organized labour in the post-war years. This fight was led by Hans Böckler of the Deutscher Gewerkschaftsbund (DGB) and the steelworkers union (IG Metall)—they were fully supported by the IG Bergbau although it could not legitimately claim that co-determination had been its main priority at this time.

The Allied acceptance of the Federal government's compromise proposals had represented the conclusion of attempts to create a totally new organizational foundation for the West German *montan*-industry. Further reorganization would still occur, resulting from arrangements made up to that point, but no major changes in principle or method of application emerged thereafter. In a diplomatic exchange of notes between the

Federal government and the three Allied Foreign Ministers on 14 May 1952 the question of settling the outstanding matter of *montan*-company ownership was finally resolved in favour of private shareholders. So the two primary objectives of the Allies, which they had pursued steadfastly, yet not always with consistency, had been undermined and completely thwarted. First, the severing of coal and steel interests did not succeed and, secondly, it had proved impossible to prevent new or reconstituted iron and steel companies reverting to those private owners who had both controlled the old trusts and played a major role in sustaining Nazi dominance. The Foreign Ministers' agreement permitted, by means of the exchange of shares, former owners to re-emerge as major shareholders in new and existing companies. Ruhr coal-mining, as is accurately reflected in Schunder's words, was victorious in maintaining 'its traditional independence'.[14] It was not so much an independence of other industries, or even of state influence, but an industry-wide independence or right to organize on a joint basis, so that it could regulate its affairs according to its own sense of priorities which were inseparable from those of the iron and steel industry.[15] These two united and integrated industries seemed powerful and influential enough, not least through their very cohesion, to organize themselves appropriately to fulfil the tasks set by the national imperatives of industrial proficiency and export success.

Politically, West Germany was constructed on a decentralized basis, with reconstruction beginning at local government level, gradually building up to the federal structure created in 1949. Fundamental economic questions, in particular matters regarding heavy industry, had been settled on an international or supranational basis. However, it was necessary in the 1950s for a form of national government to evolve, effectively exercising a unitary authority which only a particular nation-state can possess and acting as an indispensable politico-economic focus for German interests. The precise status of this new state would depend upon its success in creating both economic stability and a functioning political democracy. Steps had been taken to ensure that, within the former sphere, heavy industry would not pose a threat to democracy; how effective those steps had been could only be established with the passage of time.

1952–1958: The False Dawn

The High Authority of the ECSC was equipped with considerable interventionist powers. For example, Article 57 decrees that 'the High

Authority operates, above all, indirectly, i.e. by intervening in the sphere of pricing and trade policy';[16] yet to control prices is, to a certain extent, to control competition. The conscious intention to direct and channel competition confirms the appropriateness of Hallstein's description of the ECSC's system as one of 'organized competition',[17] through a supranational public body, an internationalized state. The DKBL was wound up in 1953, an event anticipated by leading figures in the Ruhr mining industry who had already laid the foundations for new joint organizations in conscious continuity of the tradition of the Mining Association. As early as 15 May 1952 the new employers' association was formed: the Unternehmensverband Ruhrbergbau (Association of Ruhr Mining Companies). A common organization for undertaking technical and scientific research, the Steinkohlenbergbauverein (Hard Coal Mining Association), was formed in Essen on 8 December 1952. The Ruhr Mining Companies Association was linked organizationally with the Wirtschaftsvereinigung Bergbau in Bonn/Bad Godesberg, the latter pressure group being an umbrella organization for all conceivable types of mining enterprise throughout the Federal Republic. A round table was established to discuss basic matters of joint interest with mine workers' representatives.[18] In practice, once again, and well within ten years of the end of the war, a comprehensive, all-embracing network of employers and professional organizations existed, representing an intimate interlocking of interests, unified by the imperishable *Gemeinschaftsgedanke*, a form of consciousness of industry-wide solidarity which was sufficiently invincible to withstand the shocks of Allied interference and the competitive market doctrines of the neo-liberals. This co-operative propensity for administrative coordination evinced a bureaucratic rationality consistent with the traditions of Germany's scientific, industrial culture.

This highly integrated network of employers' organizations treated the need to overcome alleged *Entflechtungsschäden* (the damage of decartelization) as a high priority. The German authorities did not accept an iota of decartelization beyond that unavoidable and continually strived for amendment and modification.[19] But by 1956 another issue demanded attention in the form of the fundamental tension between the ECSC's market philosophy, the main thrust of which was to abolish cartelistic practices, and, on the other hand, the jointly organized selling arrangements of Ruhr coal. This incompatibility put the functioning of the whole community in question. A plan of reorganization was submitted to the High Authority and, with amendments, was endorsed, coming into effect on 1 April 1956. The six selling companies were reduced to three,

three sets of two companies amalgamating.[20] The removal of the High Authority's imposition of maximum prices on Ruhr coal (the last in the ECSC to be freed) occurred on 1 April 1956 at the same time as the reorganization of the selling agencies came into effect.

The Federal government's power to set prices had been transferred to the High Authority with the creation of the ECSC; now that the latter had relinquished controls prices were free, subject only to the market as modified and filtered by the selling agencies. However, the Federal government was convinced that price increases of such a basic commodity would have seriously deleterious effects on price levels throughout the economy. Consequently, in February 1956, in order to pre-empt probable price rises, the Federal government took two major steps to ease cost pressures in the industry. It reduced the employers' contribution to the miners' welfare fund from 14.5 per cent to 8 per cent, the state covering the 6.5 per cent shortfall; and it also authorized the payment of a miners' bonus (*Bergmannsprämie*) to all underground workers, 2.50 DM per shift for face workers and 1.25 DM for other underground workers, this to be funded by direct offset against a mining company's income tax liability. Its effect would be to increase net monthly earnings by 50 DM. This measure, plus the 1.77 DM (per ton) reduction in contribution to miners' welfare insurance borne by government, and a 2 DM price increase on 1 April 1956, eased the industry's cost burden by 4.84 DM per ton of turnover.

The High Authority decreed that this bonus payment was incompatible with the ECSC Treaty; however, the Federal government simply ignored the finding and passed the relevant Act. Thereupon the High Authority offered to negotiate on the matter and the bonus was finally allowed on the understanding that the recently introduced government contribution to the miners' welfare fund would cease on 31 March 1958. The event was of outstanding importance in two respects. First, a German government had emerged with sufficient confidence and authority to defy and where necessary disregard the High Authority over a vital national interest; secondly, it revealed that a Conservative government, apparently whole-heartedly committed to neo-liberal economic doctrines, had no intention whatsoever of enforcing the discipline of the market, as expressed in the rigours of price competition, on what was still considered perhaps West Germany's most important industry.

The influence of public authorities on actual price levels was obviously no new phenomenon, with vital implications for investment, especially if profits were kept low. The bulk of investment in the mining industry had become self-financed since the Currency Reform, mostly through depreciation allowances rather than profits. The industry benefited by 173 million DM from the Investment Assistance Act of January 1952 and in 1956 (retrospectively valid for 1955), the whole basis of assessment of depreciation allowances/tax liabilities for underground workings was substantially modified in coal-mining's favour, being extended to surface installations from July 1957.[21] These accelerated depreciation allowances were to apply to existing collieries until 31 December 1965 and till the end of 1970 for new ones. This form of indirect public subsidy was to compensate for inadequate capital investment from private sources, repelled by controlled prices and therefore reduced profits. The price of coal had always been (at least, since 1919) a political price, not a market price, and was to remain so for the indefinite future. Although Ruhr coal prices were de-controlled in 1956, they had not become market prices. Not only did the Ruhr selling agencies continue to adopt a unitary pricing policy, more importantly, coal prices became subject to a gentlemen's agreement between the industry and Federal government, with the practical effect that prices could not be raised without government approval. In practice, the Federal Economics Ministry continued to maintain a very sharp supervision of coal prices.[22]

By the second half of the 1950s coal still seemed to be in every respect king. The government's Annual Report for 1956 asserted that:

The Federal Government has therefore introduced a series of measures which, partly in the long term and partly immediately, are to guarantee a permanent increase in coal production ... [the] aim of these production measures is to increase annual output by 30 million tons by 1965, to 160 million tons.[23]

The same year, on 19 October, Alfred Wimmelmann, who was Chairman of the Ruhr Mining Companies Association and the Hard Coal Mining Association, announced at a coal conference: 'Accordingly, it is anticipated that in the next twenty years the output of West German coal mining will increase by 40 million tons. That means a yearly increase of around 2 million tons.'[24] Such optimistic views were also shared by reputable academics.[25] Given growing domestic demand, and export commitments to other ECSC countries entailed in the Treaty, greater recourse had to be made to imports from third countries.

In 1955, for the first time in Germany's industrial history, the country imported more energy than it exported, and by 1957 the energy deficit on the balance of payments amounted to 1.1 billion DM. Although, after 1955, West Germany became the largest continental importer of

American coal (16 million tons in 1957), the greater part of the deficit was accounted for by oil, which displaced hard coal and lignite, whose share in total domestic energy consumption fell from 91 per cent in 1950 to 86 per cent in 1957. In the mid-1950s, it was heating oil in particular which was increasing its market share by leaps and bounds. In the late 1950s German coal was to be caught in a vice-like squeeze through competition from cheap American coal and foreign oil. But in 1956, in the face of an energy shortage, the Federal German government actually encouraged domestic coal consumers to meet their excess requirements from overseas. To this end the maximum period of validity for long-term delivery contracts for American coal imports was extended from 18 months to 36 months, as only in this way could the cheapest prices and freight rates be negotiated.[26] By the beginning of 1958 the sea freight rates for American coal reached an all-time low, $3 per ton—with an American pit-head price for steam coal in mid-1957 of $6.27 compared with $15.74 for the Ruhr equivalent, the whole ECSC area market for Ruhr coal was threatened. German coal importers expected government action at any moment to restrict these American imports and contracted 5 million tons in February 1958, and 12 million tons in March 1958. At the peak of this hectic activity, no less than 44 million tons of such coal was contracted for the following three years.

By 1957, West Germany was in the anomolous trading situation of exporting 24 million tons of coal and importing 22 million tons; 18.4 million tons of the exports consisted of coal claimed by other ECSC customers, allocated to them by the High Authority on the non-discriminatory basis of equal distribution plans based on Article III of the ECSC Treaty. Nevertheless, it could be argued that the German *montan*-industry had annexed the ECSC, in that the Ruhr's sense of community had perhaps become an important part of the European Community.[27]

Notes

1. A. Bauer, 'Die Bedeutung des Steinkohlenbergbaus für die konjunkturelle Entwicklung der westdeutschen Wirtschaft', Ph.D. thesis, Hochschule für Wirtschafts- und Sozialwissenschaften, Nuremberg, 1957, 7.
2. See G. Boldt, *Staat und Bergbau* (Munich: C. H. Beck, 1950), 105.
3. Ibid., 'under no circumstances will ever be possible to regain ownership'.
4. See p. 67.

5. H. Kost, 'Die Tätigkeit der Deutschen Kohlenbergbau-Leitung—Schlußbericht', *Glückauf*, 90 (1954), 89.
6. For a thorough account of the circumstances leading up to this significant dispute, see U. Borsdorf, 'Speck oder Sozialisierung? Produktionssteigerungskampagnen im Ruhrbergbau 1945', in U. Borsdorf and H. Mommsen (eds.), *Glück auf, Kameraden! Die Bergarbeiter und ihre Organizationen in Deutschland* (Cologne: Bund-Verlag, 1979), 345–88.
7. Decree No. 112 of the British Military Government and Decree No. 19 of the American Military Government regarding the Organisation of the German Economy, reprinted in Boldt, *Staat und Bergbau*, 137–8.
8. Kost, like Heinrich Dinkelbach, Chairman of the Board of Steel Trustees, had been a *Wehrwirtschaftsfürer* during the Third Reich.
9. Quoted in W. Albert, 'Die Wiedereingliederung der westdeutschen Kohlenwirtschaft in die Marktwirtschaft', Ph.D. thesis, Bonn, 1961, 32.
10. A decision interpreted in some quarters as a deliberate ploy, at American instigation, to thwart indigenous attempts to nationalize or socialize domestic coal production. See typically P. Schaaf, *Ruhrbergbau und Sozialdemokratie* (Marburg: Verlag Arbeiterbewegung und Geisteswissenschaft, 1978), 43.
11. It could be argued that the creation of the IAR was very much a solution to appease the French; according to Parker, 'the control powers held by the Allies in Germany were not yielded up to it in any measure whatsoever' (W. N. Parker, 'Fuel Supply and Industrial Strength: A Study of the Conditions Governing the Output and Distribution of Ruhr Coal in the Late 1920s', Ph.D. thesis, Harvard University, 1951, app. 5, 42).
12. Reserve powers were awarded to the Federal Economics Minister to ensure supplies in an emergency for important customers like the Federal Railways, electricity and gas works, iron and steel, etc.
13. In the summer of 1951 the Allied ban on coal-processing (other than coke), which had been in force since 1945, was finally lifted (the St Petersburg Agreement of Nov. 1949 had stopped the dismantling of plant, for reparations purposes, of the by-products works).
14. 'Seine hergebrachte Selbständigkeit' (F. Schunder, *Tradition und Fortschritt/Hundert Jahre Gemeinschaftsarbeit im Ruhrbergbau* (Stuttgart: W. Kohlhammer, 1959), 270).
15. The regrouping 'and reorganization of the German syndicates and cartels seems sometimes to be going on even within the various German counterparts to the Allied control agencies set up by the Allies to plan and carry the re-organisation out' (Parker, 'Fuel, Supply and Industrial Strength', app. III, 22).
16. Bauer, 'Die Bedeutung des Steinkohlenbergbaus für die konjunkturelle Entwicklung', E. 13. But a defining feature of a market arrangement is a freely operating price mechanism; however, German coal prices were directly regulated by the High Authority until 1956. Also, the ending of rationing

referred to for 1950 was, in fact, short-lived: the Korea crisis led in the autumn of 1950 to the reintroduction of *Lieferpläne* (i.e. a form of rationing) which continued to apply until 31 Mar. 1958.
17. i.e. 'organisierte Konkurrenz'; quoted in Schunder, *Tradition und Fortschritt*, 280; 'organized competition' was also precisely what characterized West Germany's so-called social market economy.
18. A substitute industry-wide organization (in the absence of socialization or its equivalent). There was also a joint employers–union office to deal with questions of bonus payments: Gedingeinspektion/Gedingekommission. On 8 Mar. 1954, a statistical unit was formed, representing all mining regions: Statistik der Kohlenwirtschaft e.V.
19. 'Even while decartelization was in progress the Ruhr *montan*-industry was already striving for changes in the measures taken' (Schunder, *Tradition und Fortschritt*, 282). See also nn. 5 and 15.
20. They were not unified on the basis of geographical considerations but mainly accordingly to the intention that each association should offer a similar balance or assortment of the relevant coal types—this was considered the best way to guarantee security of employment for miners. Further joint organizations founded at this time were: the Normenausschuß to establish common terminology and standards; a common finance and accounting agency, the Ruhrkohle-Treuhandgesellschaft mbH; and the Ruhrkohlenberatung.
21. Of capital invested, the relevant depreciation allowances amounted to 50% for moveables and 30% for immoveables over a period of four years of asset life (this varied noticeably from standard practice which generally required a minimum asset life of six years; and the level of allowance was virtually unique before similar support became available for industry in the zonal border areas).
22. An inquiry by the High Authority into the industry's cost/price structure for 1954/5 yielded the conclusion that on average the Ruhr mines were losing 3.80 DM per ton (the loss range spanning 3 to 5 DM)—this discovery was all the more revealing because it explicitly included refined products which hitherto, despite being small in quantity (chemicals, etc.), had been very profitable sources of income.
23. Quoted in Albert, 'Die Wiedereingliederung der westdeutschen Kohlenwirtschaft', 47.
24. Quoted in G. Gebhardt, *Ruhrbergbau/Geschichte, Aufbau und Verflechtung seiner Gesellschaften und Organizationen* (Essen: Glückauf, 1957), 134.
25. See typically Bauer, 'Die Bedeutung des Steinkohlenbergbaus', 38: 'It is not to be expected that the market for coal will be limited by competing products.'
26. Coal imports rose from 8.4 million tons in 1954 to 31.6 million tons in 1957, of which 16 million tons was of American origin (simultaneously, much cheaper Ruhr coal, up to 30 DM per ton cheaper, was being transported to

Italy). At the same time coal stocks held by coal consumers increased dramatically from 7.3 million tons in 1954 to 26.3 million tons in 1958, primarily due to hoarding.
27. 'Ihr Gemeinschaftsdenken ist zu einem wichtigen Teil der Europäischen Gemeinschaft geworden' (Schunder, *Tradition und Fortschritt*, 299).

5

Coal in Acute Crisis, 1958–1969

Introduction

The Storm Breaks

The winter of 1957–8 brought a totally unforeseen slump in the coal market. The general slow-down in economic growth particularly affected the highly susceptible iron and steel industry, which sharply reduced its demand for coke while the exceptionally mild winter reduced general demand for coal. Many consumers were bound to take American coal because of the recently signed long-term contracts and the availability of cheap American coal within the wider ECSC market also substantially reduced demand, especially since West European consumers were not obliged to accept coal which German suppliers had hitherto been bound to deliver under delivery plans which came to an end on 31 March 1958. These factors combined first with the existing trend of rationalization in industry, reducing the specific demand for energy, particularly coal-based electricity, and secondly with the growth of different sources of energy, particularly heating oil, in private households and increasingly in industry.

The industry had to respond by piling up pit-head stocks, at the rate of approximately 1 million tons per month, reaching a total of 13.1 million tons by the end of the year. Yet at the beginning of 1958, just as the reserve stocks were already building up, the Federal Railway had introduced a special new tariff cheapening the cost of transport of American coal to south Germany. Inevitably, the industry also responded by cancelling shifts, no less than 2.7 million for the whole of the year, with mineworkers having to forgo wages amounting to 58 million DM. Such developments, associated with the possibility of widespread redundancies in the older pits of the southern Ruhr, and highly concentrated in particular areas, no doubt created the climate which promoted a 21,000 drop in the labour force. It was precisely such developments as these which engaged the interest of the wider public, in much the same way as the great strikes of the pre-war years. Between 1957 and 1958 the consumption of coal in the energy sector fell by 2.5 million tons; the rest of the domestic market declined by a further 11.5 million tons (and iron and steel by 1.3 million tons). The industry responded to the massive fall in demand by a minuscule reduction in production of 0.6 million tons, exporting an additional 2 million tons with special government assistance, the remainder of the excess output being stockpiled.

So a crisis had well and truly arrived; how would the industry and the workers react? How, indeed, would the High Authority and the Federal German government act, as hitherto their basic approaches were barely distinguishable in practice? For both had exhorted the industry to increase production, the former urging as recently as spring 1957 that output targets of a further 40 million tons be established. Although neither institution pursued a coherent and integrated energy policy, they were each preoccupied with keeping coal prices as low as possible: in fact, energy policy was largely coal policy, and coal policy was largely coal price policy. The Federal government was obsessed with the effect on overall price levels and on general industrial activity of increases in coal prices, this basic industry being considered a pace-setter for the whole economy. The High Authority was determined to ensure that German coal prices would not rise in such a way as to hinder the growth of Germany's immediate neighbours. Virtually all government and ECSC measures regarding the coal industry were related directly or indirectly to the question of controlling coal price levels and to the consequences thereof. High Authority and government together had prevented the German coal industry from exploiting a seller's market, would they now attempt to induce the industry to cut prices and, if necessary, production, the only effective response to competition in a buyer's market?[1]

The First Coal Crisis, 1958–1959

The violent turn-round in the energy situation, with its harrowing implications for German coal, had caught all interested parties entirely by surprise. Now that the coal monopoly within the energy sector was broken, would the need for public control disappear? If not, where would the main source of public intervention be focused, on the High Authority, the Federal German government, the state government of North-Rhine Westphalia, or even the local authorities? The main addressee of demands for assistance and protection had to be the Federal government,[2] both in its own right and as prospective advocate within the relevant counsels and councils of the European Communities (the creation of the EEC did not affect the ECSC's competence in the *montan*-sector). The High

Authority did not generally undertake independent initiatives; nearly all responses by public authorities were either Federal government measures, or joint ones between the High Authority and the latter. The German coal employers themselves often made their own particular forms of response contingent upon appropriate initial action by government.

However, despite the severity of the crisis,[3] the government was generally slow to respond and its reactions displayed all the classic features of disjointed incrementalism. Its first step was to exhort coal importers (generally purchasing agencies associated with German mining companies!) to exercise import discipline—whatever the effects of such exhortation, it did not prevent the first large-scale cancellation of shifts (*Feierschichten*), commencing on 23 February 1958. By the end of April, the Federal Chancellor was holding the first formal discussions with trade union leaders from the IG Bergbau, the first of the three major consultations that year. It was an expression of the serious nature of the situation that the Federal Chancellor became intimately involved in the crisis at an early stage, although official government spokesmen continued to play down the seriousness of it. By July pit-head coal stocks had attained the record level established back in the peak crisis year of 1932. The first tangible action by government was to reintroduce shortened coal import contracts to eighteen months—not that this acquired much practical consequence, because from 4 September onwards no new import licences were issued, authorization was required for new deals on existing licences, and no further permissions were granted.

On 14 October 1958 the Council of Ministers resolved to invoke Article 95 (1) of the ECSC Treaty and introduce special measures to support the financing of pit-head coal stocks. Seven million dollars was made available to coal companies from 31 October 1958 to stockpile, on a monthly basis, coal produced over and above thirty-five days net production. This money did not have to be repaid if the national governments matched the outlay, which duly occurred (the available funds became exhausted by the end of June 1960). This tinkering with symptoms rather than causes was exposed within a mere two months when the new chairman of the coal employers, Burckhardt, confirmed on 10 December 1958 that the industry was contemplating closures entailing production cuts of some 10–12 million tons annually, with the likelihood of 100,000 miners being made redundant. The only response of significance from the government was to make, from 20 December, heavy and medium grades of heating oil subject to turnover tax from which they had hitherto been exempt—but a tax of 4 per cent was hardly likely to have a considerable impact.

On 16 February 1959, the most important measure to assist the German coal industry was introduced: the imposition of a coal import duty of 20 DM per ton of coal imported over and above a basic quota of 5 million tons (which was designated to satisfy the traditional requirements of the North Sea ports). The tariff was introduced on the formal initiative of the High Authority and was projected to last merely until the end of the year; however, the tariff was renewed at regular intervals thereafter. This measure, perhaps more than any other, expressed the unreserved and uninhibited instinct of the High Authority to support and further coal protectionism. After having advocated massive expansion of coal output for so long it had no conception of how to manage a massive rundown of the coal industry on a community-wide basis. With the habits of interventionism well established, coal protectionism seemed the easier administrative option, especially with individual member governments, like that of Belgium, for example, being determined to support their industry in any way they thought appropriate.

Following this most effective of measures, a step was announced which was to prove the most abortive of all, the coal–oil cartel, inaugurated on 17 February 1959. The agreement, fostered by the government, was struck between the three largest mining companies and five major oil companies. The latter promised to undertake no measures before the end of December 1959 to expand the sales of heavy heating oils at the expense of solid fuels; but, more importantly, they undertook not to sell heavy heating oil at a price below the prevailing world level before the end of December 1960. The cartel price of 87.80 DM per ton was easily undercut by outsiders who could sell the same grade for between 58 DM and 72 DM—accordingly, during the short period of the agreement (merely until mid-August) the outsiders increased their share of the market from 10 per cent to 20 per cent.[4]

On 1 May 1959 the five-day week for miners, which had been agreed on 1 January, came into effect, but, unlike in Britain and France which had had the five-day week for some time, the shift length was temporarily increased from $7\frac{1}{2}$ hours to 8 hours. The industry anticipated by this measure a painless cut in production of 4–5 million tons annually. On 18 June 1959, the German Coal Mining Emergency Community was formed in Essen (Notgemeinschaft Deutscher Kohlenbergbau). One task was to organize the industry's reaction in the most contested markets; its primary concern, however, was to arrange the cancellation of existing coal import contracts and licences with third countries so that German coal companies could step in to fill the gap. To compensate for the cancellation of 8.7 million tons of American coal already contracted, and to compensate for the withdrawal of the import licences for 14.5 million tons of coal, 300 million DM had to be expended; 250 million DM was provided by a state-guaranteed loan (free of charges) arranged through a consortium led by the Rhenish Giro Centre in Düsseldorf. The charge made of the consumers was the originally contracted American price, which had been higher than that for Ruhr coal and had been kept high by the import duty. The German coal industry had made this whole deal dependent on the maintenance of the coal import duty. The incident represents a fascinating example of state–private industry co-operation. The principles of the whole undertaking were worked out jointly; the industry itself provided the organization for the operation, the state provided an indirect subsidy by not charging for its service in guaranteeing the appropriate loan, and further public offices in the form of the publicly owned banks acted as the relevant financiers. The industry's capacity for cohesive solidarity obviated the necessity for direct state interference.

The Crisis Peaks

Although the steps to combat the problem of foreign coal were progressing well, there was no evidence that the substitution of coal by various grades of heating oil was slowing down. With the overall situation continuing to deteriorate the miners' union warned of the probability of protest marches and indeed, on 6 September 1959, the first demonstration in the Ruhr by miners took place. A week later, on 12 September, the SPD demanded the appointment of a Federal Coal Agent (*Bundesbeauftragter*) for Coal Mining.[5] On 16 September 1959, the government finally announced the introduction of a heating oil tax which was scheduled to run until the end of December 1962 but was not due to come into force until April 1960. A tax of 30 DM per ton was initially planned but this was strongly contested by the oil industry and lengthy negotiations finally yielded a rate of 25 DM per ton for heavy grades of heating oil and 10 DM per ton for light grades, to be levied between 1 May 1960 and 30 April 1963. The tax yield was not intended simply to augment state revenues; it was specifically reserved to finance measures to ease social hardship for miners.

The Labour Factor

In the meantime the crisis seemed to be taking its natural course, with pit-head coal stocks by the end of September reaching 17,062,000 tons, from a mere 423,000 tons in October 1957. Between January 1958 and September 1959 the number of underground workers declined from 345,447 to 295,290, with output per shift increasing substantially at the same time from 1,605 kg. to 1,912 kg. Over the same period no less than 6 million individual shifts had been cancelled, with a shortfall in wages for the miners concerned of 130 million DM. In this climate the largest post-war demonstration in Germany took place in Bonn on 26 September 1959, with 60,000 miners participating in a silent procession with black flags flying. Two days later details were announced of the agreement between the High Authority and the Federal government on measures to assist the coal-mining industry. The next day the Cabinet accepted the Draft Bill for the heating oil tax, and, of immediate relevance for the miners, a hardship fund of 75 million DM was established to compensate miners for cancelled shifts, with payment to begin from the beginning of December.

In fact, a total of 150 million DM was expended to compensate miners for lost wages between 1 February 1958 and 30 September 1959, half being borne by the High Authority, half by the Federal government. The government declaration of 16 September 1959 also contained five further provisions to assist miners in finding alternative employment:

1. If a miner obtained a new job outside his normal place of work or residence he could claim travel and removal expenses and a compensatory payment if he had to maintain two households;
2. He received a wage equalization payment if his new wage was less than that previously obtained in mining.
3. He could receive a grant to finance retraining if he would otherwise have to accept lower paid work without retraining.
4. He was entitled to a waiting payment (*Wartegeld*), if immediate alternative employment was not available.
5. Miners in receipt of miners' insurance benefits were to receive a redundancy payment to encourage them to leave the industry altogether.

The financing of all these social measures, totalling 800 million DM, was to be met from the heating oil tax. Never since the onset of the industrial revolution in Germany had miners received such massive state support (and these measures were by no means to be the last)—and no other group of German workers, either before or since, has ever received state assistance on such a scale.[6]

The sheer size of the West German labour force in coal-mining was a sufficient factor in itself to make it a potent force in industrial politics. In

1955, just one year before the peak year in post-war output, the industry employed 523,488 workers (excluding the Saar), 8 per cent of all West German industrial workers. Although not possessing a record of industrial militancy, they were highly concentrated geographically, mainly in the Ruhr, and had demonstrated a resolute determination to defend their rights in 1951 in the dispute with employers and the government over co-determination. They had displayed an organized solidarity which contrasted sharply with the divisions characteristic of the pre-war years. Coal-mining was the third largest industry in terms of manpower employed, behind mechanical engineering and textiles.

Generally Hesitant Responses

The German coal employers had certainly not acted with any greater sense of urgency than the government in combating the crisis. It was not until 21 January 1960 that the industry formed the Ruhr Mining Action Community (Aktionsgemeinschaft Ruhrbergbau) which included all the Ruhr mining companies and which was intended to manage the closing down of uneconomic pits. This was another industry-wide organization created to deal with a problem facing the whole industry. The Action Community compensated those companies which had to close down pits by granting them a payment financed by a levy on the industry as a whole; the contraction of the Ruhr coal industry thus appeared to be being borne equally by all participants. Once again, collective solidarity permitted an orderly and organized response to a new situation. One explanation of the industry's hesitant response may be offered by a narrow management perspective since the industry had previously been insulated from competition from other energy forms. Competition had hitherto been understood primarily in technical rather than commercial terms, it was very difficult for senior management, mainly engineers by training, to appreciate fully the need for an active sales policy. It could also be argued that the many restrictions imposed on the German coal industry by the ECSC hindered a flexible marketing policy although after 1958 such a complaint would no longer be valid.

Another factor inhibiting the coal industry's response to the dramatic, new situation related to the industry's self-perception of its own boundaries. Despite the unmistakable *esprit de corps* which pervaded the whole industry, particularly in its upper echelons, outside interests had long possessed a major stake. Before the crisis actually broke Ruhr mining interests had already acquired a major interest in oil refining. The BV-Aral AG, a joint Ruhr mining undertaking, was producing around 3 million tons of oil products annually and had become one of the leading liquid fuel concerns. The owner of the largest West German oil refinery at that time was the Gelsenbergbenzin, a subsidiary company of West Germany's largest coal-mining company, the Gelsenkirchener Bergwerk(s) AG. Scholven-Chemie of Gelsenkirchen, which worked in close association with BV-Aral, was owned by the publicly owned Hibernia Mining Company. The Graf Bismarck Colliery was a subsidiary company of the Deutsche Erdöl AG. If one includes the close co-operation between the Mobil AG Hamburg (with its refinery in Bremen) and the Gelsenkirchener Bergwerk(s) AG then there was an unmistakable overlap of heating oil and coal interests in no less than six of the eleven leading mineral oil companies in Germany, and they owned half its refining capacity (the output of heating oil in refinery output doubled between 1955 and 1959; from 30 per cent to 60 per cent). The failure of the coal industry to appreciate soon enough the emergence of oil as a direct competitor rather than as an additional alternative seems to have been more a case of not wanting to see rather than not being able to. In view of this overlap of coal and oil interests the prevailing arrangements no doubt seemed quite satisfactory to both the interested parties. There can be little doubt that far more decisive action was taken to fend off American coal than to restrict oil consumption, with the consequence that: 'The slimming down of West German hard coal mining . . . has . . . already virtually come to a halt.'[7]

Were the measures taken by the government and by the industry itself likely to promote a permanent and satisfactory resolution of West Germany's structural coal crisis? Given the long history of state intervention in the coal industry, given the increased involvement generated by the present crisis, and in view of the general efficiency of the industry on any objective comparison with other Western European producers, government measures would be the most decisive in shaping the industry's future. But they would not be able to promote a permanent resolution. The government had no control over two of the three main structural features contributing to coal's decline: decreasing specific consumption, and the trend towards processed or refined energies (*Edelenergien*). Similarly, the government had only slight control over the third factor, the availability of oil in general and heating oil in particular; it could not possibly deny German consumers, least of all industrial consumers, access to a fuel becoming freely available to Germany's industrial competitors. It was not fully appreciated that it was the rise of oil, especially heating oil, which was the central problem for coal.[8] If independent experts like Schaal and Albert hopelessly overestimated the ability of German coal to become and remain competitive,[9] and thereby totally underestimated the ability of oil to conquer the energy market, then it is not surprising that government officials and the coal industry's own authorities also completely misread coal's opportunity and oil's potential. A consensus had emerged based on the conviction that a slimmed-down coal industry would be viable, with minimal state assistance.

However, certain physical realities had to be taken into account, as geological conditions were very adverse. A very high input of capital and manpower was required because of the great depth of Ruhr pits, relatively frequent faulting, impurity, and the relative thinness of seams. No less than 44 per cent of the Ruhr's coal reserves were contained in seams with a moderate to steep slope, making mechanization correspondingly more expensive. But it was the considerable depth of German pits which put them at such a disadvantage. In 1953 the average pit-shaft depth was 750 metres, compared with 400 metres in France and only 60 metres in the United States. With the progressive exhaustion of seams, the deepest pit-shaft was becoming deeper by five to six metres per year. It was extremely difficult to conceive how coal-mining could succeed in cutting costs when by 1960 one-quarter of all coal mined was obtained at a depth of 1,000 metres. Moreover, coal-mining is a very labour-intensive industry, yet the level of wages is not dictated by productivity levels. The dangerous and dirty nature of work underground has determined that miners receive amongst the highest wages of all industrial wage-earners; without such top wages miners simply cannot be recruited, and even then they are recruited with difficulty. The industry was compelled to pay high wages irrespective of economic performance and had little control over an irresistible, upward trend which was being set by manufacturing industry with its high rates of productivity.

The scope for cutting costs was negligible; it was beyond the industry's ability to cut costs in absolute terms and it was also beyond its power to influence relative costs. Where were cost reductions of the desired magnitude to come from? From excessive profits? Not even the most vigorous concentration on the most profitable seams would permit any meaningful competition from Ruhr coal.[10] And it was the total failure of policy-makers to face these harsh facts of German coal's prospects and to base policy on a realistic assessment of the consequences that sowed the seeds of the next crisis. Coal could only survive with state assistance, probably of a most substantial kind.

The Real Nature of the Crisis

But what precisely had been the nature of the crisis? Whatever it was, the government and the coal employers had reacted tardily. The miners themselves did not indulge in significant protest until the crisis had been underway for eighteen months, and even then there were no strikes; there was no cause, very few miners had actually been dismissed. Older miners had been induced to leave with the equivalent of a redundancy payment; the most productive miners, in the twenty to forty years age range, had mostly transferred from uneconomic pits to more profitable ones; other miners had simply found different jobs, with or without publicly financed retraining. So many ex-miners had been able to find alternative employment in the resurgent economy that already by the end of 1960 the employers were complaining that too many miners were abandoning the industry!

Perhaps the real nature of the crisis was, in a sense, psychological. A turning-point in German industrial history had been reached. The transition had occurred fairly suddenly not least because the change-over from coal to oil, initiated at the end of the 1920s, had been long deferred. This delayed adaptation appeared at first to be progressing relatively smoothly, despite the major exception of the recent crisis, largely because the economy as a whole was expanding. Should a similar situation reoccur during a recession, then such a problem might not be resolved so easily.

That the end of an era had come was confirmed by two major developments which largely escaped attention and comment at the time. In 1958, the High Authority had loosened its control. It permitted Ruhr coal suppliers to sign long-term contracts, for more than one year, within the ECSC; they were also permitted to adapt prices to match offers from other ECSC and third countries. The *Verbot des Eintretens in die fremden Preise* (ban on the seller bearing part of the buyer's transport costs) was also lifted and, in addition, the Ruhr coal companies were allowed to grant rebates not shown in the official price lists. In short, by permitting cartelistic pricing policies the whole *raison d'être* of the ECSC had been abolished. The second development came during the crisis of 1958–9 when the freeing of *Werksselbstverbrauch* took place. Cokeries were released from their obligation to deliver coke to particular iron and steel works. The smelters had never been under a similar obligation to receive coke and they were now indicating both to independent and tied cokeries that the iron and steel companies had no intention of being handicapped by buying expensive coke based on high-cost Ruhr coal. This decision was of momentous historical import. For more than sixty years iron and steel interests had progressively acquired control of the coal industry to suit their own ends; now German coal and coke was likely to prove a chronic liability and therefore steps were taken to escape this potentially oppressive burden.

Now that the coal industry was, in a sense, on its own, with its excessively high cost structure and apparently with a basic inclination 'of reducing its own adjustment in the same proportion as imported coal is excluded',[11] its medium- to long-term prospects were sombre indeed.

The Interregnum, 1960–1965

Apparent Stabilization

During the early 1960s, the West German coal industry seemed to stabilize. Production settled at around 140 million tons per annum (1959: 141.7 million tons; 1960: 142.3 million tons; 1961: 142.7 million tons; 1962: 141.1 million tons; 1963: 142.1 million tons; 1964: 142.2 million tons) and coal stocks dropped rapidly to 4 million tons by the end of 1963. The primary cause of this satisfactory development was the boom in the iron and steel industry although a series of cold winters, particularly that of 1962–3, also helped sustain coal demand. Considerable progress in rationalization, reflected in substantial improvements in labour productivity, was a major contribution towards containing costs. The industry seemed to be enjoying the full benefit of protection from American coal competition. In this situation, the government felt confident enough, under continuing pressure from the industry (particularly from the employers), to offer what was interpreted as a support-guarantee for 140 million tons of annual production. Thus Ludwig Erhard declared in the Bundestag on 16 May 1962 that the Federal government intended 'to guide economic policy such that coal-mining can maintain a market of 140 million tons'.[12] Despite an unambiguous qualification added ten months later,[13] this was in fact interpreted by the industry as a production/sales guarantee which generated a belief that the state would give financial support for this output level, irrespective of the industry's own performance and of further changes in the energy market.[14]

Oil continued, however, its onward march, progressively taking an even larger share of the energy market, the heating oil tax in particular being largely borne by the oil companies themselves and therefore having no significant impact on price movements and demand.[15] Not surprisingly, in 1963, for the first time, German hard coal supplied less than 50 per cent of the domestic energy market.[16] As oil inexorably increased its share of the energy market, the oil companies came to represent an increasingly influential pressure group. But the distinction between a coal interest and an oil interest is by no means unambiguous. Thus the German consortium to explore for oil and gas in the North Sea was headed by the Ruhrgas AG, the Ruhr Coal Gas Company, and, by the mid-1960s, about 35 per cent of German oil refining capacity was directly or indirectly controlled by the coal industry.

The German coal industry had managed in the meantime to sustain its efforts at increasing productive efficiency and between 1958 and 1964 productivity increased by 49 per cent, no less than 9 per cent higher than that for industry generally. However, heating oil prices continued to fall, by a remarkable 30 per cent, between October 1963 and July 1964, to 70 DM per ton. In these circumstances, with oil reserves being discovered at twice the rate of the increase in oil consumption, the outlook for the German coal industry was appearing increasingly hopeless and the signs of a new crisis were unmistakably emerging. Although 1964 coal output remained unchanged over 1963, the amount actually marketed was 9.7 million tons less at 138.2 million tons. And the tell-tale rise in pit-head coal stocks was substantial, doubling the amount for 1963, at 7,784,000 tons, with the trend markedly upwards.

An Interim Review

In 1965, the storm clouds began to gather. By 30 June 1965, pit-head stocks of coal and coke had already reached 15.2 million tons, 10.7 per cent of total production for 1964. The ECSC had manifestly become redundant, the Treaty was out of date and in need of substantial revision, and there was no prospect within the foreseeable future of the development of Community-wide energy policies which would be of immediate relevance for coal.

At best, although *Land* governments and the ECSC could play supporting roles, coal policy was primarily a national/Federal responsibility. The Federal government's measures were mainly concentrated in four major areas: (*a*) social measures, intended to prevent severe disadvantage for miners arising from the structural transformation of mining; (*b*) measures granting the mining industry a protection, permitting desirable adaptation; (*c*) measures supporting increased rationalization in the industry; and (*d*) measures to stabilize the coal market in compensation for the inelastic nature of coal production.

The very comprehensive and extensive measures of support for the coal industry from 1957 represented such a massive programme of assistance that it made a mockery of the government's proclaimed antipathy towards state intervention. But despite a massive and unprecedented level of support for the coal industry central government had failed to develop a coherent, consistent, and convincing energy policy. Measures were put into effect but were not conceived as components of an integrated strategy. In anticipation of the Law to Further Rationalization in Hard-Coal Mining the government initiated an advance action programme of financial incentives for pit closures; mines to be closed would receive 12.50 DM per ton of coal production foregone. The Act created the Rationalization Association as a Federal public-law organ. It consisted not of public officials but of private individuals who were involved in coal-mining. The main tasks of the Association were to improve productive equipment and production processes and to facilitate rationalization on an inter-mine basis. In effect, a comprehensive and detailed state framework was provided for a semi-state organization to fulfil the purpose of furthering an efficient coal industry. The Association was expected to make a major contribution to the already substantial progress in rationalization, as the remarkable increase in productivity had made the West German coal industry the most efficient in Europe.[17]

Given the new, key position of electricity production in hard-coal consumption, the most important measure introduced by the government at this time to assist coal was the Act to Further Coal Consumption in Power Stations passed on 23 June 1965.[18] It was anticipated that during the period concerned more than 150 million tons of coal would be used than could otherwise have been expected. This particular kind of government support would cost 1.2 billion DM in forgone tax revenue (900 million DM surtax and 300 million DM corporation tax). Were these and similar measures intended to remedy an alleged market failure? The antidote for the deficiencies of an ill-functioning or badly distorted market and the unsystematic and contradictory efforts of government was to be planning, an undefined concept around which a non-official consensus was unmistakably crystallizing. In this some employers were almost more emphatic than labour, a significant development as the whole wider discussion of planning in general and subsidies in particular was not confined to energy and coal where the debate was most intensive. The German social market economy was reaching a crucial stage in its evolution and the developing crisis raised fundamental questions regarding the philosophy and techniques of state participation in business affairs.

In 1965, a clear recognition of the need for planning and consistent interpretations of what could be meant by the term had not yet become generally acceptable. But a unified capital–labour advocacy of the coal industry's interests, supported in so many respects by influential academic opinion, was bound to create a formidable force to influence government policy. An irresistible pressure was mounting, in favour of more comprehensive and systematic planning. But the more industrial leaders, on both sides, displayed social and political responsibility in economic affairs the less the state would be obliged to intervene. If more planning were deemed necessary in a free economy then perhaps more self-planning by the principal industrial actors would be most appropriate. The attitudes of mining labour indicated a readiness to co-operate and, equally important, the union possessed the relevant capacity to deliver.[19] A political party–interest group consensus was emerging which tended to isolate central government.[20] Now that the planning argument was additionally legitimated by an academically acceptable theoertical base,[21] the government would come under irresistible pressure to intervene more comprehensively and systematically: the eventual breaking of the economic crisis was to inaugurate a planning era.

The Federal government's decision to opt for coal protectionism was a classic illustration of non-planning: 'coal protection in the Federal Republic is the result of the combined efforts of the UVR and the IG Bergbau',[22] an *ad hoc* response of monumental proportions. The employers had much to gain by a closer association, while the union enjoyed considerable political influence, particularly within the SPD. The basic SPD stance was much more interventionist than that of the CDU or FDP, so the employers could expect more support from the SPD in government than from the CDU. A harmony of interests was emerging which would be favourable to political changes at the very apex of the West German state. This pressure was reinforced by the twin spectres of mass unemployment and political radicalization: 'the conduct of the Federal government towards mining was characterized above all by the fear of the social consequences of restricting production.'[23] Erhard had stated: 'the unequivocal aim of all the measures taken by the Federal government today is to avoid mass redundancies.'[24]

Hastening Decline and Imminent Collapse

By 1966, coal production had, however, slumped to 126 million tons, 16 million tons less than in 1964 and coal sales plummeted to 117.9 million tons, 30 million tons less than in 1963. It seemed as though the bottom was falling out of the coal-market. Falling production and increased rationalization had seen the labour force nearly halved between 1958 and 1966. It seemed that however much the industry rationalized it simply could not cut costs. Improvements only succeeded in containing the rise in costs, they could avail nothing against the cost of American hard coal, down 23 per cent and heavy grade heating oil, down 61 per cent. The former's cif price was 12 per cent cheaper and the latter's untaxed ex-refinery price 46 per cent cheaper than domestic coal. Neither government nor the industry's own measures could make German coal competitive.

With the whole economy moving into a major recession, by far the most serious since the Currency Reform of 1948, the German coal industry was possibly on the edge of complete collapse. It was no longer in a position to save itself, only drastic state action could offer the prospect of any hope at all. But the existing government was politically discredited, only a fundamental political change would permit the thoroughgoing and effective intervention required to rescue the coal industry. An emergency programme was required which could not await the emergence of a comprehensive and integrated government energy policy. The only way to cut through the Gordian knot of difficulties would be the advent of a government which was both unashamedly interventionist and was sufficiently committed to coal to give it the necessary priority. The SPD met these requirements on both counts, both because of its Keynesian interpretation of the social market economy, and because of the key position of the IG Bergbau in the power structure of the German labour movement. With the advent of the Grand Coalition, and its Social Democratic Economics Minister, Karl Schiller, the era of non-planning was at an end.

The Second Coal Crisis, 1966–1969

No supranational foundation for coherent national energy policies within the EEC/ECSC yet existed. By the onset of 1967, the union of the executives of the EEC, ECSC, and Euratom had still not occurred. The merger was to materialize later in the year, but

in coal producing countries. Each largely pursued its own policies . . . (the) existence of European Communities produced a modicum of co-operation among members. The ECSC maintained a number of programmes to alleviate the crisis. Proposals for co-ordinated policy to deal more systematically with the problems were discussed and regularly rejected.[25]

COAL IN ACUTE CRISIS, 1958–1969

The enduring coal crisis had revealed the ECSC as a limited organization unable to cope with coal's problems. As late as 1970 Gordon was also able to refer to 'the long futile efforts of the European Communities to achieve a co-ordinated policy'.[26] Conflicts of interest at supranational level, particularly between those countries which were major energy producers as well as consumers and those which were virtually energy consumers only, were to prove insuperable obstacles to the establishment of a viable Western European energy policy.

Two Simultaneous Crises: Coal and the National Emergency

In the meantime, the government had continued to implement individual measures on the basis of the usual *ad hoc* approach. Of outstanding importance was the passing of the Second Power Act on 30 June 1966: 'Gesetz zur Sicherung des Steinkohleneinsatzes in der Elektrizitätswirtschaft'; it came into effect the following day. Its main objective was to ensure that Community (German) coal would provide approximately 50 per cent of domestically produced electricity by 1970. This Act followed only a year after the First Power Act which had proved deficient. The new Act was intended to compensate electricity companies for the additional costs of building and running coal-fired power stations over and above the original tax-saving.[27] But would the Act succeed? Heating oil prices continued to fall, pit closures continued unabated, so the prospect of meeting the target without further subsidies seemed quite remote.

German industry itself did not passively await government action, it was capable of generating its own initiatives, one of the most prominent being that which created the Action Community of German Coal Regions (Aktionsgemeinschaft Deutscher Steinkohlenreviere). With Ruhr coal production in 1966 falling to 76 million tons, a means of managing cuts in production and the ensuing pit closures was required such that redundancies would be compatible with the ability of local economies to absorb the additional labour, particularly vital in traditional mining areas; clearly, new industries had to be attracted. Consequently, the BDI invited industrial and government representatives to a first meeting on 27 April 1966. The two main objectives were, first, to undertake pit closures, made inevitable by declining demand, over and above those envisaged by the Rationalization Association, and, secondly, to improve the economic structure of mining areas by attracting new industries. An Action Community private company was formed, with government financial support. It offered a closure grant (premium) of 15 DM per ton to cover continuing charges such as pensions, miners' coal, and subsidence compensation. The company could claim back the money from the government. In order to ensure the availability of land for new factories the premiums were made conditional upon the sale of all mining property, not just the collieries, to the company; this was also a way of obviating land speculation. By 1968, the Action Community had not proved in practice to have been as active as had been hoped. Not only did the severe recession inhibit smooth functioning of the scheme, the ongoing negotiations for unitary mining companies indirectly inhibited agreement, because of the need to avoid prejudicing those negotiations and possible future arrangements for additional closures.[28]

One arrangement introduced at about this time which not only proved to be successful but of historic dimensions was the introduction of a coking coal subsidy: *die Kokskohlenbeihilfe*. This assistance was absolutely vital for one very simple reason, the German steel industry was in danger of departing to coastal sites in other ECSC countries. There was nothing new in German steel companies moving to the North Sea coast (it had begun about sixty years before), but the Federal Republic was the only Continental country employing coal tariffs and quotas against American coal. This put other ECSC producers like Holland and Italy at a distinct advantage with direct access to American coal and foreign ores, quite apart from the fact that the new steel plant on coastal sites was up to date and more efficient. The German and Belgian steel industries were particularly alarmed by their deteriorating prospects. They put considerable pressure on the High Authority for appropriate steps to be taken and when it became apparent that the German and French governments were prepared to apply national solutions a supranational agreement (Decision No. 1/67) was obtained which envisaged subsidizing Community coking coal down to the equivalent of the American price. The agreement had become essential as a means of preventing the malaise in the coal industry from infecting iron and steel, as Schiller had warned could happen (in the Bundestag on 30 June 1966). The agreements on coal utilization in electricity generation and the steel industry guaranteed a minimum market for German coal in the medium term of 80 million tons.

The Grand Coalition, the Three-Phase Programme, and the Coal Adaptation Act

By the end of 1966, coal consumption was down 12.2 million tons on 1965, down by 19.7 per cent; producers' coal stocks had risen to 20 million tons. The new government, which had been formed during the winter months of November–December 1966, the Grand Coalition, came into office to face the Federal Republic's first major economic set-back since 1948. The crisis in the coal industry was both a cause and a symptom of the acute national recession, with sharply falling sales and investment, rapidly rising unemployment, and increasing balance of payments difficulties. This was accompanied by a tense political atmosphere, with the rise of right-wing extremism (NPD), and the rise of left-wing and often violent extra-parliamentary opposition, the APO, led by the SDS.

A formidable series of measures for assistance to the coal industry were already in existence before the two major new approaches were undertaken by the Grand Coalition. They included import quotas and coal tariffs; freight subsidies for coal transport; heating oil tax; self-restraint by the oil companies regarding the marketing of heating oil; financial assistance for rationalization and pit closures; support for coal-fired power stations; subsidies for coking coal and coke for the iron and steel industry; assistance in the construction of coal-fired heating stations; a very comprehensive compensation system for redundant miners and those on short time. By the time Schiller was in a position to develop new initiatives this virtually unprecedented support programme had totally failed to enable the German coal industry to maintain a constant output level let alone a stable share of the energy market. By 1966 50.9 per cent of all primary energy was derived from oil, only 32.3 per cent from coal. With the continuing availability of cheap oil and the recession-induced drop in demand, a second coal crisis had emerged with a vengeance.

Schiller's first step was to inaugurate a special Concerted Action for Mining. A series of meetings culminated in the presentation of a Three-Phase Programme for the adaptation and regeneration of the German hard-coal industry on 20 March 1967. Adaptation entailed reducing coal supply to match fallen demand, apparently a prerequisite for integrating the industry into the market economy; state measures to assist this transition were not intended to last longer than eighteen months. The phases envisaged were: a preparatory stage until the end of 1967; an adaptation phase until the beginning of 1969; and a stabilization phase from 1969 until 1970.

The most important intentions were embodied in the Coal Adaptation Law and in the proposals for company reorganization. The Bill for the Coal Adaptation Law was passed by the Cabinet on 24 May 1967 and presented to the Bundestag in November. Formally, its two primary objectives were: first, to induce colliery companies to realign productive capacity according to market requirements; and, secondly, to concentrate production on the most efficient and profitable pits. More specifically, in the Bundestag debate of 8 November, Schiller summarized the intentions of the Act in terms of a threefold purpose: reorganization of German mining companies; the creation of a comprehensive social plan (*Gesamtsozialplan*); and the creation of a structural plan for new industrial installations in mining areas. The Act became law on 15 May 1968.

The most outstanding feature of the Act was the creation of a Federal Agent for (Hard) Coal-Mining and (Hard) Coal-Mining areas, the *Bundesbeauftragter*, and the extraordinary powers granted to him. It was primarily his task, certainly not directly that of the market, to further the adaptation of the German mining industry to developments in the energy sector. The most important basis of his power regarded decisions on the allocation of state financial assistance to particular enterprises: if they did not comply with his recommendations, such support could be withdrawn. This ultimate sanction represented the stick which could be employed if previous government carrots failed to act as sufficient incentive for compliance. The key role allotted to the Federal Coal Agent was a very explicit admission of the central and indispensable part which was continuing to be played by the state in the German coal industry.

The introduction of the Coal Adaptation Law and its proposals regarding coal company restructuring occurred amidst a depressed and depressing industrial climate, the mood in the Ruhr being one of acute anxiety:

The discussion of the various plans and the energy policy of the Grand Coalition was conducted in 1967 in the tensely laden atmosphere of an almost despairing Ruhr population . . . It was not just a crisis following the recession but a catastrophe threatening . . . The economic trend was still downward.[29]

The announcement of further planned pit closures, for example, Graf Bismarck, Hansa, and Pluto, seemed to confirm that the Ruhr coal industry was collapsing like a house of cards. It was against such a background that the many plans for saving the coal industry were presented, in most cases by employers' representatives. In the early 1960s, the first major suggestion for a reorganization of German coal-mining, on the basis of a private unitary company (*privatwirtschaftliche Einheitsgesellschaft*), had been advocated by Werner Söhngen, Chairman of the Board of Management of the Rheinische Stahlwerke (Rheinstahl). His early initiative was of great symbolic importance: it confirmed the readiness of steel interests to divest themselves of what had now become their coal liability, it also laid a marker for a solution on the basis of a single company.

COAL IN ACUTE CRISIS, 1958-1969

By the time the government's own proposals appeared, a clear consensus had emerged on a number of key features of coal policy: the creation of some form of unitary company (*Einheitsgesellschaft*), and substantial state involvement. The government's Coal Adaptation Bill contained no reference to a unitary company, but the reference to 'optimal company unit' was amended during the passage through the Bundestag by 'especially a combined company' ('insbesondere eine Gesamtgesellschaft'), on the initiative of the SPD. But it had been the mining union's proposals for a unitary company, issued on October 1966, well before the new government developed its Three-Phase Programme, which first really anchored the concept in the public consciousness. Ever since the war the union's demands regarding a total, industry-wide reorganization had centred on various kinds of *Vergesellschaftung* or *Sozialisierung* (socialization), mainly in the form of *Gemeineigentum* (common ownership). Such formal terms had largely been abandoned by 1964.[30] Five years after the Godesberg Programme of the SPD, the ideological demand for some form of state ownership of the industry did not seem consistent with social democracy's pragmatic embrace of the market economy.

Over the ensuing months the union became intimately involved in negotiations with interested parties regarding the reorganization of the industry. It refined and modified its stance and in October 1967 published its own White Paper: Economic and Social Stability through Restructuring (Hard) Coal-Mining. Whatever the merits and demerits of all the union's specific proposals, there can be little doubt that in the final agreement the union's view of the new company's structure, both in regard to its apex and the seven subsidiary companies, prevailed.

The first public presentation of the Rheinstahl Plan occurred in July 1967, in an article published in the *Frankfurter Allgemeine*. The Plan stressed the necessity both for government energy policy measures and self-help measures by the coal-mining industry, the latter being dependent on the former. The proposal envisaged the leasing of collieries to one or two operating companies; this referred to the actual production facilities and not the land itself, and excluded, for example, the transfer of power production. The lease payments paid to the mining companies as lessors would amount to 360 million DM per annum and be guaranteed by the state. After twenty years the mining facilities were to become the property of the operating companies. The practical effect of the recommendations would be to cut overall German coal production to 90 million tons, of which 77.2 million tons would be provided by the Ruhr (taking the 1966 output of 103.4 million tons as the latter's baseline). The plan represented the wishes of about two-thirds of the mining companies, those owned by steel interests.

The Plan was generally very badly received. By October 1967 new press reports indicated considerable evolution in the Plan, with a commitment to a unitary company, a reduction in the state guarantee to 4.9 billion DM, and a more compromising attitude over the transfer of non-mining facilities, but the critics were not to be mollified and by early 1968 the Plan seemed to be dead. In fact, the opposite was the case, behind the scenes it continued to form the basis of discussion and negotiation. How could it be otherwise, with such influential backing? Namely from Hermann Abs of the Deutsche Bank; President of the BDI, Fritz Berg; the head of Klöckner, Dr Günther Henle; President of the Central Office of the German Chambers of Industry and Commerce, Dr Ernst Georg Schneider; and Hans-Günther Sohl, Chairman of the Thyssen Steel Company, all of whom, with the exception of Berg, possessed their own interests in coal.

Yet three outstanding characteristics typified the whole discussion concerning German coal company restructuring: first, not a single plan recommended what would conventionally be recognized as a market solution, along the lines, for instance, of the Allies' former proposals, based on Law No. 27, envisaging at least ten competing unitary companies; every solution, in effect, uninhibitedly advocated the creation of a horizontal monopoly. Secondly, most proposals, including the more controversial ones, emanated from spokesmen associated directly or indirectly with the steel industry; and thirdly, all the plans envisaged state support, usually of a substantial nature.

The Special Position of Organized Labour

Apparently, the fate of the coal industry was to be decided primarily on the basis of a deal between steel interests and government.[31] However, by the mid-1960s there had been a socio-political transformation in the status and influence of the German labour movement in general and of the mineworkers' union in particular. The union now occupied a key position in the whole structure of coal-mining's politico-industrial framework and consequently was in a position materially to influence the outcome of negotiations.

The IG Bergbau was unique in several respects. It was the only mining union, a tremendous step forward compared with the crippling divisions characteristic of the pre-1933 period. The union was also exceptional within modern German unionism in its high level of union membership, which was not only well above the average for German industry in general.[32] with the highest level of membership, but the percentage of enrolled membership actually increased during the crisis years (which was most of the time), in direct contrast to the normal experience of declining industries. Given the high degree of integration and cohesion of the union, its determination to defend itself, as amply demonstrated in the early years of the Republic in its resolute defence of parity co-determination, and the increased militancy evidenced by the recent Saar strike, the union was, in any conventional sense, a force to be reckoned with, Not that the union had to threaten the use of industrial muscle to assert its interests. By sharing management responsibility at the highest level and in particular by its equal representation on the supervisory boards of mining companies, it not only had access to relevant information, it could continually ensure the presentation of the union view in company boardrooms. The labour representatives had, in effect, become employers—not that they conceived their role in those terms. So when differences of opinion emerged it was by no means necessary to adopt an immediate course of confrontation. Not only would that have contradicted the practice of German trade unionism in the post-war period and the tradition of mining unions themselves, but generally such an approach was unnecessary. On some fundamental issues there was an identity of interest and the institution of parity co-determination was to prove a means of cementing employer and labour co-ordination, not unexpectedly at the expense of the consumer. Yet the German mineworkers' union was already at the centre of power. It was not unusual for German mineworkers' leaders to be members of Federal and state parliaments and even to become energy spokesmen of the SPD parliamentary parties. This was exemplified by Walter Arendt and Adolf Schmidt at Federal level and by Heinz Kegel and Heinz-Werner Meyer at state level in North-Rhine Westphalia. By acquiring power and influence within the senior echelons of the SPD hierarchy, union spokesmen were in a very strong position to advance miners' interests when the SPD was in power.[33]

Whatever the Basic Law may decree about the separation of formal rights and responsibilities of political parties and interest groups, the IG Bergbau seems to have become fully integrated into the political wing of the German labour movement and this belies somewhat the alleged political neutrality of German unions. The smooth and matter-of-fact integration of German mineworkers into the management of private mining companies, the responsible manner in which they manage their union affairs, and the political involvement of the union in parliament symbolizes not only to what extent German trade unions have become integrated into modern German society, but how they have become an official pillar of the West German state itself. The overlapping, interlocking nature of the IG Bergbau's incorporation into different social institutions makes a mockery of conventional pluralistic notions of modern industrial society.[34] Seen in this light, despite the serious prospect of widespread redundancies of mineworkers during the Second Coal Crisis, with the inevitable implications for the economies and social fabric of mining areas, the institutional context for defending mineworkers' interests was by no means unfavourable. And given the political necessity not to provide raw material for the agitation of right- and left-wing extremists (as embodied by the rise of the NPD and the APO), paradoxically the outlook for job prospects and related social benefits was strangely favourable for the coalminers at this time.

The Final Agreements

The first formal consultations of the Economics Minister with the Group of Five of the Rheinstahl Plan had occurred on 19 July 1967. By 4 June 1968, the first stage of negotiations, after a long series of bipartite and tripartite meetings, was concluded with the final Basic Agreement (*Grundsatzabkommen*), also known as the Bonn Paper (*Bonner Papier*). It had been agreed between representatives of the Federal Economics and Finance Ministries and the relevant Ministries of the state government of North-Rhine Westphalia, mining employers, and the IG Bergbau, and concerned the procedures for establishing a combined company for the Ruhr. It was also agreed that the Total Adaptation Programme should be effected by mid-1971. By 29 July, the Federal Agent, Dr Woratz (a senior civil servant), had been appointed. In the meantime, the Coal Act had come into force on 19 May.

On 8 October, discussions were initiated with the European Commission regarding Community authorization of the Ruhr combined company. On 19 November, the Federal Agent issued the report on his investigation. At long last, on 27 November, the Ruhrkohle AG was officially founded, but only as a pre-company (*Vorgesellschaft*), with an initial basic capital of 10 million DM, paid in by twenty mining companies representing 73 per cent of output. In fact, the second and final round of negotiations was not really successfully completed until 18 July 1969 when the Economics Minister, on behalf of the Federal government, signed the Basic Contract (*Grundvertrag*) with the Ruhrkohle AG (possessing a supervisory board of 21 members), the original coal companies (*Altgesellschaften*), and the IG Bergbau. Even then, only 85 per cent of Ruhr coal production was incorporated in the company. The final date for inclusion within the main agreement expired on 15 August. Eventually, all remaining companies were constrained to join, with one exception, August Viktoria, owned by BASF. Lucrative deals were made by hanging on for as long as possible, an opportunity created by the cyclical upsurge in the economy.

In November, the seven constituent companies of the Ruhrkohle AG were created and authorization was received from the European Commission. On the last day of that month the full agreement (*Vertragswerk*) officially came into effect. On 1 December 1969, the last agreements with the original mining companies were signed, with mines and miners officially being transferred to the new company. On 1 January 1970, the Board of Management of the Ruhrkohle AG finally assumed complete responsibility for all aspects of the company's trading. It had taken nearly three years to effect this particular solution, an indication of just how complex the reconciliation of conflicting interests had proved to be. The main interests of government had proved fairly straightforward: to restrict its financial commitment to the reasonable level of 3.3 billion DM, plus more than 1 billion DM of flanking measures. The union was primarily, but by no means exclusively, concerned with securing employment prospects and the social wage of miners. The main source of tension was in conflicting interests amongst the owners, the main bone of contention being the historical combine (*Verbund*) connections.

But even the slow progress in the restructuring of German mining companies would not have been attained unless the Three-Phase Plan and the Coal Adaptation Act had created an adequate foundation for securing the material position of the individual miner and for giving special assistance to the economic development of mining communities. These joint aims were achieved by the establishment of a total social plan, a *Gesamtsozialplan*, drawing together all the many miners' benefits in one coherent scheme, and by comprehensive measures to stimulate new investment in the Ruhr in order to create additional, alternative employment.

Important though the measures to protect miners were, from the point of view of regional development they were only short-term palliatives. Continued investment was required to generate new employment. On 2 June 1966, the government of North-Rhine Westphalia had already issued guidelines for the award of 100 million DM to coal-mining districts, the fund to be used to subsidize interest payments of new industries moving into areas affected by pit closures. In 1967, 150 million DM was made available from the European Recovery Programme special fund,[35] and made a contribution towards the Three-Phase Programme developed for the Saar, assisting the acquisition of buildings and plant. The Coal Adaptation Act itself envisaged 10 per cent investment grants.[36] The government's estimates for the 1968 budget, issued on 13 September 1967, envisaged spending 1,051 million DM on coal-mining, compared with 916 million DM in 1967 and 460 million DM in 1966.[37] On 14 March 1968, the Minister-President of North-Rhine Westphalia, Kühn, was able to announce the Ruhr Development Plan, with a planned expenditure of 8.4 billion DM between 1968 and 1973. Most of these measures were to be seen in the context of a new government emphasis on structural policy generally, emphasizing the innovatory influence of the SPD within the Grand Coalition.

Prospects of Success

But, despite this very comprehensive attack on the coal crisis, within the context of measures to combat the national recession, would the objective be achieved, the regeneration of the West German coal industry as a competitive industry operating at a lower level of production? Experts at the time were divided,[38] but a dispassionate reading of the evidence yielded a negative answer. The survival of the West German coal industry was unmistakably a political problem, either the industry would disappear entirely, or it would require vast levels of state assistance to maintain even a moderate output level. The readiness of government to provide massive financial assistance could in one light be explained by its weakness in failing to resist powerful interest groups backed by the fear of political radicalization of an alienated labour force in mining when subject to drastic unemployment. An alternative explanation might go deeper. Could it be that the collapse of the German coal industry would induce the German steel industry to emigrate?[39] Would not the dramatic demise of the *montan*-sector have such a deleterious effect on business confidence generally that Germany might suffer an economic débâcle, with frightening political consequences, paralleling the experience of the 1930s?

The trading relationship between the German steel industry and the Ruhrkohle AG was regulated in the so-called Foundry Contracts—these long-term contracts for delivery of coal (of twenty years duration) were intended to replace the severed vertical links between the steel companies and Ruhr coal-mining. But these contracts were constructed in such a way as though the previous association still existed; They were predicated on the continuation of the coking coal subsidy: 'The final arrangements did not reflect equality of status and interest, namely . . . the Ruhrkohle AG was obliged to deliver all it produced whereas the steel concerns were committed to taking coal for a maximum of three months, with a binding order only being made at the beginning of each quarter'.[40] This obligation meant that both short- and medium-term planning of production by the Ruhrkohle AG could be rendered ineffective because only the Ruhrkohle AG bore the risk if the foundries' projections of their future needs proved to be inaccurate. In other words, the steel companies dictated their own terms in practice.

An identical situation obtained in regard to relationships between Ruhr coal-mining and power generation and distribution. The Ruhrkohle AG was, however, allowed to provide much of its own electricity: it was permitted a total generating capacity of 951 megawatts, consisting of the mining industry's oldest and least efficient power stations, with outputs varying between 7 and 82 megawatts. A power station contract was signed with the thirteen parent companies whereby the Ruhrkohle AG undertook to supply them with generating coal, on the assumption that state measures of assistance would continue. The same one-sided relationship as with steel obtained, yet at least in this sphere steadier growth could reasonably be expected as the electricity industry was less vulnerable to the cyclical variations which so affected the iron and steel industry.

Interim Review: The Pattern of Interests

One of the most intriguing aspects of the second coal crisis was the pattern of interests which emerged and the mode of their expression. The least ambiguous interest was that of organized labour. Its main concern was to prevent job losses, and, where this was not possible, to establish a system of welfare benefits which minimized, if not actually eliminated, the financial hardship entailed by redundancy. Indeed, no other industry, before or since, has created such a comprehensive, effective, and generous system; not even the steel industry, which had also wrestled with major redundancies, had, by 1980, approached the standard established by the miners. Not that the miners' union had to resort to militant measures to protect its members. There was not a single strike over redundancies—and national and state civil servants paid unreserved tribute to the positive and constructive role played by the IG Bergbau, a view partly shared by the coal employers.[41]

The union of employer and labour interests was strengthened by the common threat to production and jobs created by falling demand, the two coal crises acting as catalysts in the cementing of relations. Organizational co-operation found expression primarily within the institution of parity co-determination which was unique to the *montan*-sector. But from the point of view of the employers this enforced alliance had originally been something of a shot-gun wedding, an arrangement to be endured because circumstances decreed no alternative. There can be little doubt that the employers strongly resisted this post-war innovation. However, the first coal crisis inaugurated a transformation in employers' attitudes for they came to realize that a united employer–labour front would be able to exert greater pressure on government for assistance than if the two sides of industry fought separately. Co-operation with the IG Bergbau was especially facilitated by its moderate stance and its high level of organization minimized, although it could never exclude, expressions of internal dissent enabling the union to deliver on an agreement. Indeed, the general context of labour relations in West Germany by the early 1960s was certainly favourable to such co-operation for: 'The relationship of employer and labour organizations in the Federal Republic is characterized more by friendly co-operation than by hostile confrontation.'[42]

The view of several American social scientists at the time, that the power of employers and labour unions could counter-balance one another,[43] was misleading in the German context: 'Federal German experience in recent years has shown, however, that this system does not always need to be effective. Namely, both groups can also unite at the expense of a third party, here the consumer.'[44] Quite a strong case could be established that this is precisely what happened with the creation of the Ruhrkohle AG. Although an even stronger case could be made that the taxpayer, more than the consumer, was the real victim of this unholy alliance for the main consumers, steel and electricity producers, had been judge and jury in their own cause (with government sanction).

In fact, we see the emergence of a new eternal triangle of interests between state, capital, and labour in the West German coal industry which established a possible precedent for other industries and offers a classic illustration of the following argument that:

When competition can no longer fulfil its control function then the way is opened for other methods and institutions of market control which install the complicity of labour and capital within large companies. Such paths lead indirectly to state intervention when decisions about market supply are transferred from the realm of the market into that of the company, out of the public view into the confidentiality of bodies with parity representation.[45]

The 1970s were certainly to evidence an unprecedented harmony between mining capital and labour.

The interests of capital were by no means unitary, but possible areas of conflict between coal, steel, electricity, chemical, and other interested parties were neutralized by contracts with the Ruhrkohle AG which guaranteed highly favourable arrangements. In fact, such arrangements would not have been attainable without state sanction evidenced by government signatures to the relevant agreements entailing huge financial commitments. This development does not confirm the view held by certain Marxists that the state was proving to be the mere tool of heavy industry, pumping public capital into moribund industry in order to guarantee a financial return on what was proving to be misplaced investment. The iron and steel industry was quite prepared to uproot basic production and relocate it on the coast of the Netherlands. The electricity generating industry would much rather have preferred to employ cheaper imports of coal, gas, and oil than to continue to utilize expensive German coal.

The decision to protect coal, at virtually any cost, represented a political decision resulting from the fear of the consequences of abandoning coal to the market. Without protection, the uneconomic position of German coal would mean instant collapse of the industry, with the basic iron and steel industry emigrating and massive unemployment concentrated in particular *montan*-districts. The failure of business confidence and a dramatic, rapid rise in unemployment could have created a level of social unrest which would have threatened the political stability, indeed, the very survival, of the Federal Republic. Seen in this light, the relative unanimity of the political parties is readily explainable, the stakes were far too high to permit a partisan indulgence in political gain. A new tradition of party accord had in any case emerged in the Federal Republic, but in retrospect the formation of the Grand Coalition, which was attacked in many quarters at the time as unnecessary, as just a cynical exercise in power-brokering at the expense of a vigorous parliamentary democracy, can be seen to have been justified.[46]

In terms of practical politics there was not much scope for wide differences of approach from that of the major political parties in regard to coal. Given the relatively large number of people still employed in coalmining, the many engaged in associated supplies, and the families of those concerned, a very substantial body of voters had to be considered.

Their votes had the potential not merely to influence, if not decide, who formed the government in North-Rhine Westphalia, but also who formed the government in Bonn. North-Rhine Westphalia, being the most populous Federal *Land*, was represented by approximately 120 Members of the Bundestag. It would be a very rash political party which did not appear to give convincing priority to the rescue not only of threatened mining communities but also to that basic sector of the economy which still seemed to be the foundation of German industrial society. Once again, the highly concentrated nature of the problem had made the situation potentially explosive. The crash programme to save the coal industry appeared successful in defusing the situation as the national economy experienced a remarkably speedy turn-round in its fortunes. The two oil crises of the 1970s seemed to some to have retrospectively justified the action taken regarding coal; however, the missed opportunities of the early 1960s actively to diversify the Ruhr economy were merely obscured by these later developments and the consequences were to make themselves sharply felt twenty years later.

Notes

1. It would not be strictly accurate to refer uncritically to 'energy policy' at this time—one section of the Federal government's Annual Report of 1951 refers to coal-mining, the iron industry, and the energy sector; coal was considered just one, even if by far the most important, mining activity among many others—the energy sector (*Energiewirtschaft*) did not embrace oil either, it referred exclusively to electricity and gas and throughout most of the period under review, 1940–1980s, there was no coherent body of law relating to the whole energy sector; in fact, the Power Law (*Energiewirtschaftsgesetz*) of 1936, applying just to gas and electricity, was to remain 'the' corpus of energy law. For the first time, in 1956, there was reference in the government's report to the 'area of energy' (*Energiegebiet*), implicitly involving coal, oil, and electricity; see W. Albert, 'Die Wiedereingliederung der westdeutschen Kohlenwirtschaft in die Marktwirtschaft', Ph.D. thesis, Bonn, 1961, 39, 47.
2. Constitutionally, natural resources were a Federal responsibility, according to the Basic Law, Art. 74, Section 11.
3. Cf. K. H. F. Dyson and S. Wilks, 'The Character and Economic Context of Industrial Crisis', in *Industrial Crisis: A Comparative Study of the State and Industry* (Oxford: Martin Robertson, 1983), 1–25.
4. But, most noticeable and puzzling of all, why did a government which was apparently so thoroughly committed to competition actually take the determined initiative to create precisely a cartel, of all things?

5. His primary concern, advised by a body of independent experts and representatives of the employers and the miners' union, would be to co-ordinate all measures relating to coal-mining—in particular, he (they) would be required to develop a long-term plan to guide the industry's adaptation to the new energy environment.
6. See also K. Lauschke, *Schwarze Fahnen an der Ruhr: Die Politik der IG Bergbau und Energie während der Kohlenkrise, 1958–1968* (Marburg: Verlag Arbeiterbewegung u. Gesellschaftswissenschaft, 1984), 245.
7. Albert, 'Wiedereingliederung der Kohlenwirtschaft', 135. The German hard-coal industry had become dependent on the protective tariff, enabling it to pursue its time-honoured practice of dual pricing, with German coal being sold cheaply abroad; but the high domestic price encouraged substitution, requiring maintenance of the heating oil tax: government policy was locked in a vicious circle.
8. e.g. at the time of the abortive coal–oil cartel of 1959, the equivalent Ruhr coal price, to compete with oil on the basis of equal calorific value, would have been between 40 and 50 DM, while in fact the price was 66 DM. By 1960, typical Middle Eastern production costs for one ton of oil amounted to 7.25 DM; for one ton of Ruhr coal 61 DM.
9. See P. Schaal, 'Möglichkeiten der Anpassung des Steinkohlenbergbaus an die veränderte Struktur des Energiemarktes', Ph.D. thesis, Freiburg, 1961, 127, and Albert, 'Wiedereingliederung der Kohlenwirtschaft', 140.
10. Nevertheless, for a detailed account of the high profits some coal companies were able to make in the early 1960s on the basis of government measures to stabilize the industry, see P. Schaaf, *Ruhrbergbau und Sozialdemokratie: Die Energiepolitik der Großen Koalition, 1966–1969* (Marburg: Verlag Arbeiterbewegung u. Geisteswissenschaft, 1978), 65 and 81–6.
11. Albert, 'Wiedereingliederung der Kohlenwirtschaft, 102.
12. Quoted in U. Specht, 'Die Energiepolitik der Bundesrepublik von 1948–1967', Ph.D. thesis, Albert-Ludwigs-Universität, Freiburg, 1969, 38.
13. Ibid.: 'at the same time I would make clear that the Government declaration does not offer a market guarantee for the said amount but represents a desirable aim to be achieved by the joint efforts of all concerned' (quoting from Erhard, 29, Mar. 1963, in the Bundestag).
14. By 1962, electricity production had become German coal's main consumer, taking 25% of its output; 55% of all German electricity was generated from domestically produced coal. From 1962 onwards natural gas showed the highest rates of market expansion of all energy forms; by 1964, its annual growth rate reached 44%, the main impulse coming initially from major supplies discovered off the coast of Holland. Multinational oil companies owned two-thirds of the reserves of natural gas under the North Sea and by 1965 were producing 44% of the output. See G. Krink, 'Die energiepolitischen Maßnahmen der Bundesregierung', in H. Schmidt (ed.), *Energiewirtschaft und*

Energiepolitik in Gegenwart und Zukunft (Berlin: Duncker & Humblot, 1966), 170.
15. The demand for crude oil in West Germany rose from 12 million tons in 1957 to 59 million tons in 1964, of which only 7.7 million tons was produced domestically.
16. Coal could not possibly compete, given the following developments in prices: the pit-head coal price per ton rose between 1957 and 1966 by 5.9%, from 63.29 DM to 67.00 DM; but the price of 1,000 hard-coal units of heavy grade heating oil fell from 95.20 DM to 37.36 DM (pre-tax ex-refinery price).
17. Underground OMS rose from 1,599 kg. to 2,705, putting Germany ahead of Britain and France whereas it had trailed in third place in 1957. The overall effect was to maintain total production at a constant level, i.e. increased efficiency compensated for the loss of production capacity of closed pits, but this artificial level of output could not be justified in relation to international competition: it was supply without demand.
18. The basis of the Act was a system of tax incentives (granting a 45% reserve fund, free of tax, calculated from actual construction and running costs) which would compensate owners for the extra cost of building and running coal-powered stations.
19. Moreover, mining leaders Walter Arendt and Heinz Kegel were particularly well equipped to speak for the German labour movement as a whole. Arendt was not only head of the mineworkers' union, he was an SPD MdB and energy spokesman for the parliamentary party, and also a European MP; Kegel was also a prominent member of the Düsseldorf Landtag.
20. Cf. C. Deubner, 'Change and Internationalization in Industry: Toward a Sectoral Interpretation of West German Politics', *International Organization*, 38 (1984), 519.
21. Particularly influential was Professor Hans K. Schneider, Director of the prestigious Energy Institute of Cologne University.
22. Specht, 'Energiepolitik der Bundesrepublik', 55. (This was also confirmed to the present author during private conversations in the Federal Economics Ministry, summer 1982).
23. Ibid. 38.
24. Quoted in ibid.
25. R. L. Gordon, *The Evolution of Energy Policy in Western Europe: The Reluctant Retreat from Coal* (New York: Praeger, 1970), preface, p. ix.
26. Ibid. 230.
27. It was estimated that 120 million more tons of coal would be consumed by 1971 as a result of the Act—at a total cost of 1.65 billion DM. This works out at a subsidy of 15 DM per ton. However, if one also includes the taxes forgone by the government over the same period, 3.0–3.4 billion DM, then this means a total subsidy for generating coal of 4.5–5 billion DM: this represents (on average) a subsidy of 40 DM per ton. Schaaf details how

COAL IN ACUTE CRISIS, 1958-1969   127

electricity generating companies associated with coal-mining were able to indulge in substantial profiteering, at the tax-payers' expense, by manipulation of the said Act's provisions; see Schaaf, *Ruhrbergbau und Sozialdemokratie*, 265-6.

28. Schaaf also illustrates how certain mining companies deliberately employed delaying tactics in order to win time for lucrative deals in selling off coal stocks profitably (*Haldenbestände*) as the economy picked up during 1968, and especially for sales of land and housing, with spectacular profits (ibid. 260-1).
29. F. Spiegelberg, *Energiemarkt im Wandel: Zehn Jahre Kohlenkrise an der Ruhr* (Baden-Baden: Nomos Verlagsgesellschaft, 1970), 167.
30. At the IG Bergbau's Wiesbaden conference.
31. See also Gordon, *The Evolution on Energy Policy in Western Europe*, 44-5.
32. In the early 1960s, 84.9% of miners were organized in the IG Bergbau; compared with 37.5% in the metalworking industry, 18.1% in construction, 27% in transportation, and 38.7% in leather-working.
33. A chairman of the energy group of the parliamentary party of the SPD in the 1970s, Erich Wolfram was an active member of the IG Bergbau, formerly a full-time official, and he played a prominent role in the practical development of SPD energy policy, especially when the SPD was last in office at Federal level.
34. IG Bergbau represents here only the ultimate development of what obtains generally regarding the reality of the practice of the German labour movement, with, for example, Heinz Oskar Vetter, a member of the IG Bergbau, former head of the DGB and former deputy head of the IGBE, and Eugen Loderer, former head of IG Metall, the largest metalworkers' union in world, becoming not only individual members of the SPD, but also Euro-MPs of that party.
35. The following year, a structural programme for the Ruhr-Saar Zonal Areas, costing 250 million DM from ERP funds, generated a total investment volume of 2 billion DM; see Schaaf, *Ruhrbergbau und Sozialdemokratie*, 301.
36. The Second Cyclical and Structural Programme of the Federal government, promulgated on 10 Aug. 1967, contained a *doppelter Bevölkerungsschlüssel* for coal-mining areas.
37. Not including financial support for the miners' social insurance scheme, with a planned increase from 2,750 million DM in 1967 to 3,194 million DM in 1971.
38. Spiegelberg was very optimistic, see *Energiemarkt im Wandel*, 194; a more pessimistic view was expressed in F. Burgbacher (ed.), *Ordnungsprobleme und Entwicklungstendenzen in der deutschen Energiewirtschaft* (Essen: Vulkan Verlag Dr Classen, 1967), 158. Specht exemplifies the schizophrenia evident in a number of German academics at the time, e.g. quoting negative evidence, p. 48, and coming to a positive conclusion, p. 108 ('Energiepolitik der Bundesrepublik').

39. In early 1969 it was confirmed by Cordes, a member of the Board of Management of the August Thyssen Foundry, that the steel industry had seriously considered a geographic separation of basic iron and steel production from metal processing; see Schaaf, *Ruhrbergbau und Sozialdemokratie*, 323. The former production would transfer to coastal sites in Holland and the latter to the western end of the Ruhr; this argument was expressly employed to pressurize the government.
40. K. Buddee, 'Determinanten und Entwicklungstendenzen des Absatzes der Ruhrkohle AG und die Problematik einer absatzorientierten Produktion', Ph.D. thesis, Cologne University, 1974, 79.
41. Civil servants in the relevant Economics Ministries in Bonn and Düsseldorf offered unstinting praise in interviews held with the present author in the summers of 1982 and 1983; although less enthusiastic, employer spokesman also expressed an attitude of general approval.
42. H. Schneider, *Die Interessenverbände* (Munich: Günther Olzog Verlag, 1965), 97.
43. See J. K. Galbraith's concept of 'countervailing power'.
44. Schneider, *Interessenverbände*, 155.
45. G. Schneider, 'Neue Allianzen für die freiheitliche Ordnung', in *Zwang und Grenzen der Konzentration* (Düsseldorf: Handelsblatt, 1966), 54.
46. For a representative, dismissive view of a so-called 'national emergency', see Schaaf, *Ruhrbergbau und Sozialdemokratie*, 209–10.

6

Twenty Years of Retrenchment, 1970–1990

A New Beginning for German Coal?

The Grand Coalition's major measures, the Three-Phase Programme, the Coal Adaptation Act, and the creation of the unitary company, the Ruhrkohle AG, were all intended to put West German coal production on a new and sound footing. A consistent continuation of government policy towards the industry was more or less guaranteed by the SPD remaining in power after the 1969 general election and, in particular, by the continuing presence in office of Professor Karl Schiller as Federal Economics Minister. But the prospects for the West German hard-coal industry at the beginning of the new decade could hardly have been less favourable. On the basis of no state support, the ECSC possessed, at best, a viable total output of 100 million tons.[1] More realistically, only 50 million tons could be produced profitably by the British coal industry and none on the Continent. During the period 1956–70, the share of hard coal in West German p.e.c. had declined from 71 per cent to 30 per cent. The West German economy, like that of the rest of Western Europe, now relied largely upon an oil-based energy sector. Moreover, natural gas was also competing strongly. During the ten-year period of 1960 to 1970, its share of total energy consumption within the EEC trebled to 9 per cent.

Yet, in 1970, Schiller felt justified in declaring that coal and the coal areas still had a future, as confidence in mining returned, especially in the Ruhr.[2] At that time, the industry was working to full capacity to meet a buoyant demand and stocks were at a low level. However, cost increases since 1969, particularly those arising from changing currency parities, made the medium- to long-term prospects more problematic. Success could, though, be claimed for the government's employment-generating coal programme,[3] with 118,000 new jobs being created, of which 85,000 were in North-Rhine Westphalia, and most of these in the Ruhr. But the emphasis here had been on metal and metal-processing, electricity generation, and chemicals, hardly a thoroughgoing diversification, with most jobs related to the existing basic goods sector.

The coal industry's main hope was that further restructuring and rationalization would increase productivity and thereby improve competitiveness, and also help to attract and retain the necessary skilled miners. However sharper increases in oil prices would not necessarily benefit German coal because natural gas and foreign coal would remain unaffected. Regeneration of the industry was not yet occurring and a strong case could be made that such a policy was doomed from the beginning. But it was as clear to government in the early 1970s as it had been during the late 1960s that at stake was the survival of the German *montan*-sector, the associated maintenance of general business confidence, and, above all, the securing of a still vulnerable parliamentary democracy.[4] Structural change had to be conducted at a socially acceptable speed. Without the obvious, potential seriousness of the situation, it is unlikely that continuing support for coal on a massive scale would have been acceptable to the CDU/CSU-dominated south German states where industrial progress had been facilitated by oil, not coal.

But for the time being coal's position did not appear critical and it seemed that a state, capital, and labour consensus on raw materials generally would be able to cope with likely stresses and strains. That consensus was based on wider agreement on how to manage industrial affairs, both at macro- and micro-levels. The employers accepted that it was a state responsibility to guarantee the competitive order. Economic policy would determine the security of electricity supply, and energy policy the security of the supply of coal and coke to the iron and steel industry. A joint concern for the mining work-force inspired a certain identity of position for coal employers and mining labour. The former feared that redundant miners would be unobtainable once business improved, and the latter demanded the complete removal, by all technical and organizational means, of fluctuations in the business cycle on mining employment.

Common interests dictated that co-operation, not competition, was the overriding imperative: co-operation nationally among domestic suppliers, co-operation between government and the private sector; and co-operation internationally both in terms of state–private ventures abroad, and inter-governmental co-operation. Consequently, energy policy, economic policy, and foreign policy had become inextricably interlinked. The strongest advocacy of co-operation in all its forms came from the state service: co-operation had to begin at home.[5] The concrete embodiment of practical co-operation was a form of planning, reinforced by the familiar need to reduce risk in investment.[6]

The great unspoken assumption was that planning by the state formed the prerequisite for effective planning in the economy as a whole. The early 1970s witnessed that stage in the development of the Federal Republic when planning was automatically assumed to represent the primary means of guaranteeing stable economic growth. This optimistic and confident belief was very much associated with the advancement of the SPD to the position of major coalition partner, after its election victory in 1969.[7] The role of the state in planning was not merely a passive, supporting one but, where necessary, an active, initiating one: 'Such planning presupposes strength, namely the strength to act.'[8] As long as the purposes of the planning and the means of their fulfilment were not too closely specified, the prevailing consensus could continue. It was certainly not endangered by labour demands for nationalization or any other form of social ownership; that simply was no longer an issue.

But one particularly ominous cloud was beginning to loom on coal's horizon: the rise of an articulate and sometimes militant environmental awareness. Coal's particular environmental problem was that it was much harder than other fuels to desulphurize. Sulphur dioxide emissions represented just one among many hazards associated with energy generation: 'Every part of the energy sector ... is more or less confronted with problems of environmental protection.'[9] Accordingly, the social-liberal coalition government presented proposals to parliament to implement its environment programme in October 1971.

However, during the 1970s economic growth was the overwhelming national priority. In the words of a professional politician like Kienbaum (FDP): 'One can almost say that the energy cost structure of a country is the basis for judging its future competitiveness.'[10] The seven most energy-intensive branches of industry, which accounted for 80 per cent of all industrial energy consumption, were responsible for 45 per cent of Germany's industrial exports.[11] The imperatives associated with such facts justified Economics Minister Schiller's mission to ensure that the state fulfilled its responsibility to equalize interests. The state's role was that of an active referee, with the role of the professional politician virtually indistinguishable from that played by a public administrator/civil servant. Relevant policies would not develop in a desirable direction merely as a result of an equilibrium of competing forces and interests, as in conventional pluralistic models; the result had to be fair, a moral rather than a technical evaluation, to be established by that superior arbitrator, the state. Internal unity, not the disunity of atomized competition, was essential for the external struggle, that of Germany's industrial trading relationships. German industrial supremacy required technical excellence, and a high and increasing input of energy: that was the priority which could not be endangered and would apply irrespective of technical matters of the ownership of the means of production, distribution, and exchange. It is the great continuity of the German political economy: the enduring centrality of industrial prowess.

A Third Coal Crisis?

Since its foundation, the RAG had hovered on the brink of bankruptcy.[12] The total adaptation plan (*Gesamtanpassungsplan*), envisaging an investment outlay of 2 billion DM, had appeared on 28 June 1971, but had then been superseded within a few months by a new emergency programme of October 1971, anticipating a cut in production down to 66.8 million tons and also entailing more redundancies than originally planned. The adaptation plan had turned into a straightforward closure plan.[13] By 30 May 1972, a redevelopment plan (*Sanierungsplan*) had been agreed, with annual costs over the three-year period, 1972–5, accumulating to 3.4 billion DM, conditional on speeding up the original adaptation plan closures. But, in essence, this latest form of improvisation represented another palliative, a further bridging operation before renewed action was called for: a final solution was inherently impossible. Given the low priority of German coal with German steel interests, and the low profile of the big banks, coal employers were forced into an ever closer alliance with the one ally with a shared interest—mining labour. But this joint force became ever more dependent on state support. Despite continuing advances in efficiency, a successful slimming down of mining was out of the question and collapse appeared more likely than consolidation, posing the perennial question regarding 'to what extent a further reduction in output can be justified socially and regionally and therefore be realized politically'.[14]

The initiative clearly lay with government to organize a joint solution between its own political bodies, the coal industry itself, the main customers, and the social partner, the mining union IG Bergbau. The industry certainly needed partnership. If the prevailing consensus disintegrated then the industry would collapse. Only the exercise of a corporatist prestidigitation had enabled the industry to survive so long. The government had also become, in part, a victim of its own propaganda. Having presented the protection of coal in terms of security of supply and concern for the welfare of the miner and his family it could not with any credibility suddenly retreat from that position, especially with the regular occurrence of *Land* elections. The state had already erred badly before 1958 in predicting an enduring energy shortage. A further and equally dramatic reversal would not only undermine the credibility of a particular government, it would serve to undermine the legitimacy of the democratic state. In a political culture obsessed with political legitimacy, to mishandle the coal question would represent very high risk politics indeed.

At long last, on 3 October 1973, the government presented to parliament the first attempt in the Federal Republic to provide a comprehensive, all-embracing energy policy. The First Energy Programme had been announced in advance in the new social-liberal government's Declaration of 18 January 1973. For the foreseeable future oil would meet more than 50 per cent of West Germany's energy requirements, but coal was still the main problem child (*Sorgenkind*) of the energy sector. The Energy Programme devoted as much space to coal as it did to oil, nuclear energy, natural gas, and lignite combined. After one-and-a-half decades of substantial state support, coal was meeting only 24 per cent of German energy requirements and provided only 36 per cent of electricity, despite the two Power Acts. German coal was the dearest form of energy available, with no immediate prospect of a decisive and lasting improvement in its competitive position.

The objective for coal now was one of consolidation;[15] slimming down (*Gesundschrumpfung*) was a thing of the past. Yet a most important aspect of the Energy Programme was the way it represented those two cardinal features of the West German political economy: the principle of the state and the principle of co-operation—'the state possesses a high degree of responsibility which complements that of those companies operating in the energy sector'.[16]

The Energy Programme was a commitment to an active role for the state. What was true of atomic energy was true of the whole energy sector: 'As very great expenditure and high risks are associated with these development projects state support is indispensable.'[17] Indeed, it could be argued that it was high capital expenditure and risk avoidance which had made a major contribution to attracting state interest in the mechanics of German industrialization throughout the present century. During the last fifty years state involvement had become intimate. This circumstance was totally independent of the alleged realization of a so-called social market economy. The capital requirements of German industry, which had always been paramount, represented a continuing priority whatever the socio-political regime in existence.

The First Oil Shock and its Aftermath

The social-liberal government's Energy Programme, endorsed by the Cabinet on 26 September 1973 and presented to parliament on 3 October, was rendered obsolete within weeks of its appearance. The OPEC oil price rise of 70 per cent on 16 October was followed by an increase of 130 per cent on 1 January 1974. The sea change in energy prices, which had been taking place since 1971, advanced in one sudden burst to transform the conditions of production, especially in Western industrial nations, in an irrevocable manner.

For the success of any German reaction to this new situation 'a close co-operation between raw materials producers and consumers, banks and the state will therefore be necessary'.[18] Internal cohesion was the prerequisite for external success—the Federal Republic could not afford the luxury of thoroughgoing competitive principles when foreign competition and national survival were the vital issues. Government recognized the need for a substantial state contribution and intimate co-operation. Although the former emphasized the necessity for self-help initiatives by industry itself, increasing state commitments entailed an appropriate increase in its influence; i.e. more say in the affairs of, generally private, companies, not in a dominating, domineering sense, but in a spirit of partnership. However, the desired dialogue between the two sides, state (*Administration*) and industry (*Rohstoffwirtschaft*), required appropriate machinery to provide continuity.[19] A higher degree of integration than hitherto was required of German industry, without which state involvement would not be effective. It was almost as though government were pleading for a fundamentally cartelistic approach, as if the imperatives of Germany's economic position were once again enforcing a state–industry alliance. For its part the government had already taken appropriate steps by creating an Inter-Ministerial Committee for Questions of Raw Materials. While co-operation was by now not new, it needed further development and refinement: 'Business requires a Concerted Action and the creation of a concept on a broad and representative basis.'[20] The German political economy seemed to be reverting to type. As had been the case for the last 100 years, economic success in international markets, a condition not only of national survival but of international prominence, was a sufficient competitive spur to industrial efficiency.

The oil energy crisis was to bring about casualties, one of the first being environmental protection. Another victim of the crisis, although perhaps not so immediately apparent at the time, was so-called planning.

There was already growing disillusion with the efficacy of political planning, particularly in as far as its implementation was affected by financial constraints. Before the first coal crisis every major official organization had predicted continuing energy shortage and the need to greatly expand coal production; the sudden appearance of an energy surplus had been totally unforeseen. Forward planning was intended to obviate such errors. Yet despite previous warnings from various quarters, including the major oil companies themselves, that fundamental changes were afoot in the international oil markets, the oil crisis caught all the major industrial nations of the free world, including the Federal Republic, totally unprepared. The apparent pretensions and deficiencies of capitalist planning had been completely exposed.

The outstanding feature of the government's First Revision of the Energy Programme, endorsed by the Cabinet on 23 October 1974 and published in November, was its intention to reduce substantially oil's share in total West German energy provision, in line with EEC recommendations which had appeared in May of the same year. Although the government's primary objectives had not altered, the big drop in oil's share of the market, a fall from 55 per cent in 1973 to 44 per cent in 1985, was to become one of the primary means of their fulfilment. This was justified, among other things, by a new priority for security of supply.

The oil crisis did not precipitate an expansion of the nuclear programme. The Revision's preoccupation with the contribution of nuclear energy had little to do with that crisis, more with the increasingly adverse public reaction to the said programme. Nor did the crisis lead to a renaissance of the West German coal industry. It was now accepted that foreign coalfields could produce coal much more cheaply. Any notion of reintegrating coal into the social market economy was very much a thing of the past. In fact, policy-makers were expecting a widening gap between costs and earnings and emphasizing the practical limits to the taxpayers' ability to purchase security. Indeed, the German coal industry was in danger of becoming a financial bottomless pit, absorbing investment required elsewhere in the energy sector.

Yet, despite the seriousness of these difficulties, under prevailing circumstances there was no chance of the state abandoning the industry. The government decided to create a national coal reserve of 10 million tons, to be financed by public resources.[21] But the most important step forward was clearly represented by the Third Power Act which was passed by parliament in December 1974. It envisaged the firing of 33 million tons of Community coal annually in German electricity production—the electricity utilities were not prepared to accept a higher amount, but they were now legally obliged to take the amount contracted. By compensating electricity producers for their extra costs (compared with firing oil and gas), coal-fired electricity generation became an attractive long-term proposition, thereby stabilizing this particular market.

Prelude to the Second Oil Crisis

Between 1971 and 1975 there was a fivefold increase in most posted oil prices, but this failed to stimulate coal production in the Federal Republic. Total 1975 production was 92.39 million tons. Hard coal was supplying only 24.8 per cent of the electricity generated (down from 31 per cent in 1974), having easily been surpassed by lignite, which was supplying 31.2 per cent. When the recession really began to bite in 1975, it fully exposed the extremely vulnerable position of the coal industry, not least because of its dependence on the iron and steel industry, the very industry which was reacting most sensitively to deepening recession.

More remarkable was the failure of the Third Power Act which had set 33 million tons as the official target for electricity generation. Although the electricity utilities had agreed collectively to try to consume that amount of coal annually, the Act did not bind individual utilities to take particular amounts. But more important still was the nature of the gas contracts signed back in the 1960s between the German electricity utilities and Dutch suppliers,[22] which had not been taken into account by the Act. Prompt action was required and, on 10 December 1975, new Regulations to the Act were published, raising the *Kohlepfenning* (coal penny) to 4.5 per cent of the electricity tariff. This new arrangement was expected to raise 1.4 billion DM to finance the required coal purchases for electricity generation. With all other energy forms now in abundant supply, a reduction in overall energy demand, and the coal industry's own stocks rising rapidly again to an exceptionally high level, further support measures were needed. One measure which was introduced promptly was bringing forward the establishment of a national coal reserve by one year; from the beginning of 1976, this had been seen as essential to maintain the liquidity of the coal companies.[23]

Coal's Hour at Last?

Despite serious delays in the implementation of the atomic energy programme in the late 1970s a general surfeit of energy obtained in the

Federal Republic. As control over the supply of oil in the 1970s passed increasingly to producer governments and their joint associations and the power of the multinational companies declined correspondingly, the latter turned their attention to coal. Would German coal share in this renaissance of the industry? Despite coal being 30 DM per ton cheaper than oil, there was little evidence of any meaningful substitution; energy consumers were perturbed by fluctuating currency parities, which made the difficult process of accurately gauging future energy demand all the more problematic. Demand from the steel industry was still depressed compared with 1974, while environmental constraints, particularly reflected in legal uncertainty, proved a major disadvantage.

But coal was still certain of state support. Despite the closure of 25 million tons of capacity since 1969 the Ruhrkohle AG was still the largest company in the Federal Republic exclusively providing energy, and accounted for 73 per cent of all German coal production. It was one of the largest mining companies in the world, with a total turnover in 1978 of 11 billion DM. Its capacity was 68 million tons per year, with an output of 61 million tons in 1978. Its coking capacity was 17 million tons (23.5 million tons before the onset of the steel crisis), placing it third in the world after Nippon Steel and US Steel. The RAG had 130,000 employees, one-fifth of all employees in the Ruhr (down from 186,000 in 1970). In short, the net value of its annual output was approximately 6.5 billion DM, making the company in 1977 the fourth largest in the Federal Republic, after Siemens, Daimler Benz, and Volkswagen.

An updating of the Second Energy Revision was approved by the Cabinet on 16 May 1979. The Continuing Energy Policy and Enhanced Energy Conservation document was primarily concerned with promoting energy saving in various ways. This was justified by the latest bout of substantial oil price rises, soon to be known as the Second Oil Crisis.[24] However, the most significant development regarding German coal was represented by the intention of the electricity utilities to increase their use of coal in 1979 beyond the levels originally set out in the ten-year contract with the coal industry, thus saving a further 6 per cent of previously planned oil input for electricity generation. Although investment costs of oil-based power stations were about 20 per cent less than coal-based ones, the actual fuel costs of heavy grade heating oil and of power station coal were approximately equal during the first half of the year 1979, rising above the latter during the second half of the year, making subsidies under the previous arrangements superfluous.[25]

Support for the German coal industry cost the state 6 billion DM annually. It was justified partly because coal gasification and liquefaction were supposed to possess a great future. A major joint effort of government and the coal industry was to be undertaken, on a scale twice described by Helmut Schmidt as 'unusual': 'Its realization will at the same time obtain a leading position in world markets for our investment goods industry, our mechanical engineering industry.'[26] In other words, the future of German coal was not automatically assured by the future high price of coal, it was dependent upon massive state support which could also be justified in the boost it represented for key German engineering industries, maintaining their prominent position in the world markets. Not only had such considerations been of major importance throughout the greater part of Germany's industrial history, it also indicated how far the state was prepared to go to influence the demand which certain industries were to respond to in producing their supply: the selectivity of the market was being exogenously influenced in a major way.

The optimistic government view that coal would acquire growing importance for the whole energy market in the medium and long term was shared by the coal employers, although in absolute terms no dramatic resurgence was envisaged. Compared with the industry's recent past this represented a remarkable advance. The excellence of German mining technology had already established it as a major export earner with distinct prospects for expansion. The most dramatic developments, however, which were working in coal's favour were the increases in oil prices. The increased cost of petroleum imports was 18 billion DM more than in 1978, an amount greater than the whole turnover of the German coal industry. But the argument which carried the most weight with government was the danger to social and political stability arising from endangered economic growth resulting from inadequate security of supply, against which coal was allegedly a guarantee. The industry was convinced that domestic political circumstances were satisfactory for coal policy, evidenced by the now widely accepted slogan 'Vorrang für die Kohle' ('priority for coal') and by the continuation of the 'away from oil' ('weg vom Öl') policy.

The years 1979 and 1980 were promising for the future of the German coal industry. In both years the costs of oil imports increased and in 1980 the electricity utilities signed a new agreement guaranteeing the domestic market in electricity generation until the end of the century. Meanwhile the general election of 1980 saw the re-election of the social-liberal coalition. But the government warned that subsidies for German coal were reaching financial limits and that coal and steel would have to adjust to a decline of the subsidy. There were other disadvantages for German coal. The second explosion in oil prices after 1978 caused higher costs and prices, raised unemployment, and created a virtually unprecedented balance of payments deficit.[27] The higher oil bill was removing domestic purchasing power in the order of 30 billion DM per year, resources which could have been used to support Germany's traditional basic industries. The future of the German coal industry was linked to cheap foreign coal, and much now depended on an expanding and secure supply of imported coal.

Coal in the 1980s

On 4 November 1981, the Third Revision of the Federal government's Energy Programme was inaugurated. Its main objectives were: first, to make further progress both in the economical use of energy, especially oil so reducing oil's share of p.e.c.; secondly, to continue domestic coal stabilization measures, while accepting the need for greater coal imports; thirdly, to diversify further natural gas supply; fourthly, to increase the output of nuclear power stations in the basic load category; fifthly, to increase the contribution of distance heating; and, finally, to promote a regionally balanced sourcing of oil supplies. By the end of the year the Federal and North-Rhine Westphalia governments also authorized public subsidies to assist the restructuring of the EBV coal company of the Aachen coal field.

However, the world-wide recession of the early 1980s affected the highly susceptible steel industry, resulting in a significant drop in demand for German coking coal. This decline reached crisis proportions for the coal industry by 1983 and a major 'Coal Round' (*Kohlerunde*) was inaugurated. The participants were the Federal government, the governments of North-Rhine Westphalia and the Saar, the coal employers and the union, the IGBE. They agreed that as a direct result of the fall in demand from the steel industry, total capacity in the German coal industry would have to be reduced by 10 million tons by 1989, i.e. down to 80 million tons. The consequent drop in mining employment would be met primarily by increased early retirement of miners, financed fully by the Federal and both *Land* governments (subject to the necessary budgetary constraints). It was the job guarantee policy of the authorities, either through direct guarantee or a generous redundancy scheme, which ensured a socially acceptable run-down of the mining labour force whenever the need for further entrenchment became acute.

A further step to assist the coal industry at this difficult time related to the National Coal Reserve; namely, the deferral of the repurchase obligations of the mining companies to the end of 1987. In November 1984, the EBV received further financial assistance to support ongoing restructuring. In October 1985 the Federal government and the government of North-Rhine Westphalia resolved to continue financial support for German coking coal beyond 1988 but within a clear limit and on condition of the phasing out of coking coal exports. Both steps reflected the inevitable pressure of public expenditure restraints; the second step represented a response to ECSC pressure. This further formalization of government support had been the indispensable prerequisite for the new Foundry Contract between the German coal-mining industry and the German steel industry, signed in December 1985. It was the successor to the original Foundry Contract, enshrining the steel industry's commitment to meeting its coke requirements from the domestic supplier, something not possible without requisite financial support from the state.

By the mid-1980s 5 billion DM of public aid was being made available to the industry, some 1.1 billion DM going to coking coal (38 DM per ton) and 2 billion DM from the *Kohlepfennig* (coal penny) to electricity generation. The inherited liability from previous mining operations was costing 170 million DM per annum. Social measures, including various pension benefits and the so-called miner's bonus (*Bergmannsprämie*— 10 DM per shift for underground employees), amounted to 440 million DM. Investment grants cost 370 million DM and coal research and coal processing cost 450 million DM. Hard-coal industry operations were carried out by six companies in four coal-mining areas. Some 80 per cent of the amount produced was mined by the Ruhrkohle AG, Europe's largest private coal producer. German hard coal's share in the production of electricity amounted to some 27 per cent in 1985. Nearly 90 per cent of Germany's power production was produced from domestic or quasi-domestic energy, that is, from indigenous resources of hard coal and lignite, from nuclear energy, and hydroelectric power.[28] Yet it was still a harsh fact of economic life that the German coal industry simply would not survive without government subsidies.[29] By 1986, the energy market was dominated by two major factors: the continuing slump in oil prices, due to a world-wide surplus, and the aftermath of the Chernobyl nuclear power station accident. In nominal terms, the oil price was now back to the level that preceded the Second Oil Shock of 1979/80. But the contribution made by German hard coal to securing supplies required unusually high and increasing expenditure; in the words of the Economics

Ministry. 'The Federal government is very concerned at the development in public aid to coal.'[30]

Between 1983 and 1985 alone, the Federal government, the mining *Länder*, and electricity consumers had provided around 16 billion DM for German coal (not including the grants to the miners' pension fund). By 1986, German coal was tying up 1.87 billion DM in Federal budget funds and about 2.5 billion DM from the Power Generation Fund. Foreign coal was now available at a price which was cheaper by about 100 DM per ton. Difficulties were exacerbated by the increasing costs of installing pollution control equipment, e.g. flue de-sulphurization units, in the power generating industry. Moreover, the previous consensus on nuclear power appeared to be breaking down.[31]

Hitherto, energy policy, as many other areas of government policy, had advanced on a bipartisan basis. It was a primary preoccupation of the 1986 Energy Report to re-establish non-partisan policy-making in this vital area:

The Federal government sees it as an important task for the coming legislative period to restore the consensus between the main political forces and between the Federal government and if possible all the *Länder*.[32]

But the threat to the Federal Republic's well-established corporatist policy processes was more apparent than real. The SPD did not advocate the immediate shut-down of the nuclear industry and its leaders were well aware that for an effective implementation of its policy much wider and stronger support was needed, something not likely within the immediate future. The main source of dispute centred on the need to finance German coal's immense demands. Not only was the coalition government, not least the FDP Economics Minister, Martin Bangemann, anxious to exercise vigorous financial restraint, the non-coal *Länder*, particularly the CDU/CSU states of south Germany, were also proving increasingly resistant to sharing the costs of coal support.

However, pressure on the Federal government was augmented by the new approach being adopted by the European Communities towards national subsidies, in the words of the government's own Energy Report:

On 1 July 1986 the new aid regulations came into force in the European Community. These regulations permit the continuance of German coal policy on the one hand, but on the other they also require a lasting improvement in the economics of the coal-mining industry to qualify for national aid.[33]

The government needed to justify its actions more than in the past and to be prepared to meet greater resistance. The relevant ECSC Decision

(No. 2064/86 of 30 June 1986) was primarily motivated by the implications arising from the energy surfeit. Previously, Community policy had intended its support system to ensure a certain long-term capacity to meet a proportion of energy needs from internal hard-coal sources. Now the emphasis was definitely more upon support systems to influence positively structural adaptation.

In October, the government felt compelled again to extend the contract regarding the National Coal Reserve (until 31 December 1993, with the coal companies repurchasing the stocks from 1 January 1990). In November, a further 100 million DM was granted towards the EBV's restructuring. The increasingly critical situation had forced the IGBE to publish its own proposals in July,[34] to try to influence opinion so that the priority for coal would be maintained. Not only were oil prices low, foreign coking coal was cheap and Germany's nuclear power stations were producing more electricity than initially planned. State financial support for coal was now running at over twice the level of 1984. Prospective job losses by 1995 were 25,000. By the end of 1986, the industry was employing 177,900 people. Its output was 80.3 million tons and company coal stocks had risen to 14.9 million tons. In 1970, one-quarter of output had been exported, now it was just 15 per cent of a smaller total. Particularly significant was that whereas 43 per cent of production had gone to the steel industry, now 52 per cent went into electricity supply and only 35 per cent to steel. Losses in the steel and heating markets were particularly critical, and led to a new Coal Round in December 1987, before the 1983 five-year-plan had been fully implemented.

The main new decision in the Coal Round was further to reduce capacity by 13–15 million tons by 1995. Government had already authorized public expenditure of 3.5 billion DM to support sales to the German and European steel industries and it now agreed to finance an additional 11 million tons of exports. The EBV mines in Aachen were to be run down completely. The Saar coalfield was to cut 1 million tons of output. The cost of the EBV closure and the accompanying social measures was to be met by government sources. Indeed, measures to protect and shield the miner from loss, according to the principle of *Sozialverträglichkeit* (social compatibility), was considered by the German labour movement to be exemplary.[35] An unstated assumption of these measures was that coal imports would rise.

There had never been the slightest doubt that the government would stand by the coal industry at this very critical time. Apart from miners, 200,000 jobs were at stake in small- and medium-sized suppliers to the coal industry. Between 1966 and 1986 the state of North-Rhine Westphalia alone had paid out 14 billion DM in coal support measures. In the words of the North-Rhine Westphalia Economics Minister: 'The destruction of complete mining centres and steel sites would also strike at the nerve of the whole German national economy.'[36] And in the words of the Federal President, Richard von Weizäcker: 'The whole Federal area is affected by, and involved in, efforts at structural renewal of the Ruhr whether one wishes to appreciate this or not.'[37] Nevertheless, the existing pressures on German coal continued into 1988.

EC pressure on the Federal government to reduce its subsidies to the coal industry increased considerably in 1989 to avoid a serious distortion of the internal market to be created by the end of 1992. While the EC Commission endorsed Federal government aid to electricity production, it required the Federal government to reduce such subsidies by 30 September 1989, and to establish, by 1993, a plan to restructure further German coal-mining. On 29 May 1989 the Coal Employers' Association and the mining companies formally appealed against the Decision to the European Court. The Decision was then suspended. The Federal government joined the appeal on 2 November 1989.[38]

On 28 August 1989 the Federal government and the two mining *Länder* agreed on the establishment of a commission of inquiry,[39] whose terms of reference were to create policy guidelines for a national coal policy, within the context of European energy policy; and to make recommendations on how best to arrange the continued firing of German coal in electricity generation after the Century Contract expired in 1995. Despite the uncompetitiveness of German coal production and the certainty that it would remain so in the foreseeable future, its survival was never questioned: 'the Commission pursues the objective of maintaining a viable German coal-mining industry for reasons of national and European security of supply.'[40] The primary, if implicit, purpose of the inquiry was to offer a vigorous and carefully documented restatement of national commitment to the domestic coal industry, coupled with suggestions for reform which would be sufficiently convincing to satisfy an increasingly sceptical yet determined EC Commission.

The case for the maintenance of the coal industry was based on the traditional argument of security of supply, the realization of which is unambiguously a state responsibility. This led to a surprising and controversial conclusion: as security of supply was the exclusive responsibility of national government, it followed that the one-third contribution of *Land* governments (Saar and NRW) towards financing coal subsidies became superfluous. This followed logically from the assertion that: 'efficient mining aims at economic efficiency not output targets determined by social and regional policy'. Such an argument was theoretically consistent; whether it was compatible with the realities of West German coal politics was another matter.

The second main thrust of the Report dealt with a so-called 'optimization process' (*Optimierungsprozeß*)[41]—greater efficiency through greater internal competition. German coal companies should compete more with each other and, much more importantly, operating units within them should compete with one another on cost grounds to establish which pits were more likely to be viable in the longer term. The effective implementation of such a policy would permit the establishment of a meaningful efficiency limit (*Effizienzgrenze*): an optimal total output for the industry. The Commission majority envisaged 35 million tons being fired in electricity generation in the year 2005, of a total output of 55 million; the minority envisaged only 25 million tons in electricity production and a total output of between 35–40 million. These recommendations were consistent with the ideology and principles of the so-called social market economy and also compatible with the general intention of the EC Commission to create a free internal energy market, and in particular with its desire to cut coal production costs to enable a cut in subsidies. Nevertheless, the security of supply argument meant that 'subsidies... are necessary for a long period'.[42]

The increased pressure of the EC authorities on the Federal government was not simply a logical consequence of making adequate preparations for 1992/3; the huge German subsidies, in comparison with other EC countries, compelled attention. In 1989, imported coal was 170 DM per ton cheaper than German coal, and total subsidies averaged out at 66,000 DM per employee! For all employees together, this amounted to more than the total wage bill of the whole industry—the BDI was the one major German organization to advocate a complete running down of the domestic coal industry. The test case at the heart of the dispute with the EC Commission was the financial aids to coal-based electricity generation. No action was taken during the Coal Commission inquiry. After it had reported, events in Eastern Europe and, above all, the imminent integration of the GDR further delayed negotiation on a possible agreement.[43]

The Ruhrkohle AG had managed to contain costs so that between 1984 and 1989 prices had not risen in real terms. Nevertheless, at least 5 billion DM or up to 40 per cent of turnover was represented by subsidies of various kinds. The prospects for substantially cutting cost and subsidies are remote. The renewed determination of the EC Commission to achieve meaningful reductions in state subsidies, at least in the German coal industry, does not face auspicious prospects. The German government has acquired the political influence that its economic stature has long since justified, with a fall-back position clearly delineated by the Mikat Coal Commission: 'As long as there is no agreed, supranational energy policy, German energy policy within the EC must, according to the Commission's view, have a national orientation.'[44]

The 1990s are likely to prove as eventful a decade for German coal as any previous one. Although the Century Contract was not due to be renewed until 1995, by the summer of 1990 the EC Commission had only authorized payments from the Power Fund for 1990, yet the Federal government had committed itself to coal deliveries of 40.9 million tons per annum until 1995. In 1993 the National Coal Reserve was scheduled to be eliminated and the ECSC (2064/86) authorization of other coal aids was due to expire. EC approval of the Foundry Contract was to expire in 1997. It remains to be seen whether the new slogan of optimization will prove to be any more meaningful than the earlier ones of slimming down (*Gesundschrumpfen*) and consolidation.

Notes

1. R. L. Gordon, *The Evolution of Energy Policy in Western Europe: The Reluctant Retreat from Coal* (New York: Praeger, 1970), 211–12. See also M. Backes, 'Die Entwicklung der Arbeitsproduktivität im deutschen Steinkohlenbergbau, 1956–1970', Ph.D. thesis, Rheinische Friedrich Wilhelms-Universität, Bonn, 1974, 233–4, and K. Förster, *Allgemeine Energiewirtschaft* (Berlin: Duncker & Humblot, 2nd edn., 1973), 139.
2. K. A. Schiller, *Die Kohle und die Reviere haben eine Zukunft* (Bonn: Federal Ministry of Economics, 1970), 7.
3. By 1 Mar. 1970, the Federal Coal Agent had authorized 10% investment grants, on the basis of Paragraph 32 of the Coal Act, which facilitated investment to the value of 12 billion DM; the original parent companies had also had a sum of 2 billion DM guaranteed by the government and channelled through the private banks for investment outside the coal industry, but within the Ruhr. The tax concessions financing investment thus far (by 1 Apr. 1970) amounted to 1.25 billion DM in revenue forgone.
4. This is intimated in the words of the senior social democrat Alfred Nau, regarding the coal predicament: 'The experience has taught us to avoid

anything that could lead to a crisis atmosphere and social discontentment'. (A. Nau, Introductory Remarks to a Conference held by the Friedrich Ebert Stiftung on 23/4 Nov. 1970, proceedings published in E. Anderheggen *et al.*, *Mineralische Rohstoffwirtschaft: Planung und Perspektiven* (Bonn-Bad Godesberg: Neue Gesellschaft, 1971), 8).

5. See U. Lantzke, 'Sicherung der Rohstoffversorgung der Bundesrepublik', ibid. 27.
6. K. D. Arndt, 'Probleme einer langfristigen Planung in der mineralischen Rohstoffwirtschaft in der Bundesrepublik und im Energiebereich', ibid. 17: 'Planning is to reduce uncertainty.'
7. And also much associated with the person and influence of Professor Dr Horst Ehmke, head of the Chancellor's Office.
8. Arndt, 'Probleme einer langfristigen Planung', 17.
9. J. Kruse, 'Energiewirtschaft der Bundesrepublik Deutschland im Wandel: Ausmaß und Folgen der Strukturverschiebungen', *Wirtschaftskonjunktur*, 24 (1972), 34.
10. G. Kienbaum, 'Für mehr Wettbewerb am Energiemarkt', in H. Anders, K. H. Hoffmann, and F. Voigt (eds.), *Energie: Leistungen Prognosen Alternativen* (Mannheim: Verlagsanstalt Courier, 1972).
11. Nowhere was this more apparent than in the sphere of atomic energy: 'An alternative to atomic energy, which would not lead to great economic setbacks and to the loss of our world economic position, is not in sight.' See W. Obernolte, 'Entwicklungstendenzen in der Elektrizitäts- und Gaswirtschaft und ihre Konzequenzen für Wirtschaft und Staat', ibid. 136. This was a view backed by both wings of the German labour movement; see ibid. 19 and 41.
12. Ending 1969 with losses of 200 million DM; 1970: 500 million DM; 1971: 400 million DM—an accumulated total since 1969 of 1.23 billion DM. Figures taken from P. Schaaf, *Ruhrbergbau und Sozialdemokratie: Die Energiepolitik der Großen Koalition, 1966–1969* (Marburg: Verlag Arbeiterbewegung u. Geisteswissenschaft, 1978).
13. At this time, the accident rate rose, making it the highest within the EEC; there was also a significant increase in reported illnesses.
14. H. Reintges, 'Grenzen der Rationalisierung im Steinkohlenbergbau', *Glückauf*, 109: 7 (1973), 407.
15. Federal German Government, *Die Energiepolitik der Bundesrepublik* (Bonn: Drucksache 7/1057, 3 Oct. 1973); henceforth referred to as Energy Programme (I), 16.
16. Ibid. 61.
17. Ibid. 11.
18. H. Burckhardt, 'Rohstoffversorgung und Rohstoffpolitik', *Glückauf*, 110: 1 (1974), 1–2. Burckhardt was President of the interest grouping, the Mining Association, in Bonn.
19. D. Rohwedder, 'Die Rohstoffversorgung der Bundesrepublik Deutschland', ibid. 3–7, esp. 6.

20. Ibid.
21. Two further improvements were the stepping up of government investment assistance, i.e. grants, from 160 million DM to 210 million DM, and additional help with financing the *Altlasten*.
22. The former had contracted substantial amounts at cheap prices, but were subject to a penalty payment (fine) if the agreed amount was not actually taken.
23. The effect of the arrangements in regard to gas were, in practice, to compensate the utilities for the financial penalties involved in the take-or-pay agreements with the natural gas supply companies (see n. 22). Another significant step to assist the coal industry in the vital field of electricity generation was an additional (2nd) amendment to the Third Power Act, incorporated in the Regulations of 7 May 1976, which reimbursed electricity utilities not only for the additional costs in regard to oil, but also in regard to imported coal and, above all, natural gas.
24. A crisis inaugurated at the turn of 1978/9 by the serious civil disturbances in Iran.
25. For the time being, coal's position seemed to have genuinely improved, though not in relation to natural gas where, for example, power station investment costs were between 30–50% less, depending on the size of the station.
26. Chancellor Schmidt, 'Erklärung der Bundesregierung zur Energiepolitik nach dem Europäischen Rat und dem Weltwirtschaftsgipfel', Federal Press & Information Office, *Bulletin*, 87 (July 1979), 810.
27. In the last quarter of 1980, a further very sharp rise in oil prices occurred: the war between Iran and Iraq was primarily responsible.
28. B. Braubach (BMWi), 'Coal Subsidy Policy in the Federal Republic of Germany', speech given at the Second Coal Consumers' Conference of the Spanish Association of Mining Engineers, Madrid, 6 May 1986, 6.
29. 'Without subsidies, German companies would no longer have an economic basis for existence' (ibid. 5).
30. Federal Economics Ministry, *Energy Report of the Government of the Federal Republic of Germany* (Bonn: No. 279, Sept. 1986), 50 and 51. See also ibid. 48.
31. See A. Geißler and B. Riegert, *Energiepolitik für eine lebenswerte Zukunft* (Bonn: Verlag Neue Gesellschaft, 1988), 123.
32. *Economics Ministry, Energy Report* (1986), 2.
33. Ibid. 51; it continues: 'Convincing efforts by the coal industry will be the best guarantee that the Federal government may expect the necessary international approval for its coal policy in future as well.'
34. IGBE, *Überbrückungskonzept für den deutschen Steinkohlenbergbau* (Bochum: IGBE, 1987).
35. See N. Ranft, *Vom Objekt zum Subjekt: Montan-Mitbestimmung, Sozialklima und Strukturwandel im Bergbau seit 1945* (Bonn: Bund-Verlag, 1988), 234.

36. A speech by Professor Dr Reimut Jochimsen: 'We won't let the mining areas down' ('Wir lassen die Bergbaureviere nicht im Stich!'), reprinted in *Die Zukunft der Kohle: Plädoyer für einen neuen nationalen Energiekonsens* (Düsseldorf: NRW-Government, 1987), 18.
37. Quoted ibid. The Ruhrkohle AG had also become the largest industrial trainer in the whole of the Federal Republic; see Ranft, *Vom Object zum Subjekt*, 289.
38. Somewhat reluctantly. One German argument was that significant shifts in policy were the responsibility of the Council of Ministers, not the Commission.
39. The first meeting of what came to be known as the Mikat Commission took place on 6 Oct. 1989; it was due to report in Mar. 1990.
40. See Introduction, *Coal Commission Report* (Essen: Federal Government/ Ruhrkohle AG, Mar. 1989), 4. So it comes as no surprise to find asserted in the final chapter: '[German] coal-mining will only acquire a long-term perspective when the permanent use of domestic resources is renewed as an energy policy objective' (ibid. 169). Consequently, in the final recommendations (*Grundaussagen*) prefaced to the main Report: 'the coal option as a domestic energy form must be kept open through continued mining of coal' (Point 3).
41. Recommendations, Point 4. For further details, see *Coal Commission Report*, 176. This body was also known as the Mikat Commission (see n. 39), after its chairman, Professor Paul Mikat, former CDU Minister-President of NRW.
42. Ibid., Point 5.
43. There was virtually no hard-coal mining of significance in the GDR. Between 135,000–140,000 workers were employed in lignite mining, perhaps representing a short-term asset for the IGBE, with a prospective huge influx of new members; in the medium term it was more of a liability than an asset: large-scale redundancies in that hopelessly overmanned industry were a certainty (West Germany's sizeable and efficient lignite sector only employed between 25,000 and 30,000 workers). But East German lignite workers are in a minority among the half-million prospective new members of the IGBE; yet the issue of overmanning and redundancy is still a pertinent one. An inter-union conflict also arises: between the IGBE and OTV, the public service union.
44. *Coal Commission Report*, 184.

PART III

Continuities in Sectoral Self-Governance

7

The German Tradition of Self-Regulation

Industrial Self-Regulation: Industrial *Selbstverwaltung*?

In the late 1970s, a number of German academics, foremost among them Volker Ronge, began to lay increasing stress on the significance of the concept of a 'solidaristic self-organization of the economy'. This phenomenon apparently contradicted, or at least substantially modified, accepted notions of increasing government/state intervention in capitalist economies.[1] Against a ' "nationalization" of economic processes and functions as a general necessary tendency of today',[2] he posited a form of 'quasi-politics' (*Quasi-Politik*), arguing:

> There are, considered at least as 'relatively equi-functional', pre-state alternatives to 'politicization'; the alternatives at the centre of attention here can be designated as 'solidaristic self-organization' of individual capital units.[3]

But to what extent do collective, private agreements, representing primarily employers' interests, provide a political alternative? In many ways, the state is still indispensable: it would normally regulate much business activity through an appropriate legal framework, varying in scope and detail according to the industry in question. With various industries regulating different aspects of their own affairs this necessarily reduces the necessity and responsibility for government intervention. However, final arrangements would normally have to meet state approval, so independence from the state is far from total, and could in specific cases, in practice, be very limited.

A vital question concerns the extent to which collective private agreements on an industry-wide basis, irrespective of whether they include labour or not, can be identified with the common interest (*Gemeinwohl*), the securing of which has traditionally been a central preoccupation of the state. Be that as it may, it has been unambiguously manifest in the evidence adduced regarding the West German coal industry that the latter has exemplified a truly remarkable level of industry-wide co-operation. The industry evinces the longest and probably one of the most highly developed traditions of concertation and co-ordination of employer interests in Germany.

The first example of large-scale employer collaboration occurred in 1858 with the founding of the German Mining Association (*Bergbauverein*), the oldest, for long the most prestigious, and possibly the most influential of all individual German employers' associations. The apotheosis of coal owners' integration occurred in 1893 with the establishment of the Rhenish-Westphalian Coal Syndicate, the most famous and effective cartel witnessed by the pre-First World War world. Two of the least commented upon, yet most significant, developments occurring after the First World War were the creation of Ruhrgas AG in the 1920s and of STEAG in the 1930s—they demonstrated how competing coal employers could cooperate to such an extent that they were able, of all things, to exercise communally an entrepreneurial function.

After the Second World War, Germany saw little diminution of the cartelistic propensity: throughout the 1950s the question of the Ruhr Coal Sales Associations remained a constant source of tension and negotiation within the ECSC. It would seem that the final creation of Ruhrkohle AG represented the inherent logic of the industry developing one coal company for the Ruhr, ostensibly the result of economic necessity. However, the matrix of political and social as well as economic factors had long since programmed the complete integration or unity of Ruhr coal-mining. The German coal industry appears to be a prime example of an industry capable of organizing its affairs in such a way as to make direct state involvement superfluous, at least in the form of public ownership: true self-organization in action.

In Ronge's further refinement of his argument,[4] 'self-regulation' was synonymous with the previous 'self-organization' and, more importantly, labour was included, conceived more as an ally than as an enemy of capital. Labour involvement did not require more government involvement, the opposite being seen in the trade union tendency to rely increasingly on a mutual self-help principle (*Subsidiaritätsprinzip*) and the trend of employers and unions to engage in forms of non-state politics in the field of industrial relations, particularly regarding wages and conditions. Where industry regulates itself the state in that sphere becomes redundant and a form of non-state politics occurs, what Ronge calls 'quasi-politics'. His argument also refers to the 'structural inability of a "competitive economy" to act jointly'.[5] However, the history of German capitalism has indicated that joint action in its various manifestations *has* been one of its defining characteristics, not least in the *montan*-sector. Vigorous domestic competition has invariably been eschewed in Germany, the ultimate anathema being so-called 'cut-throat' competition. Again, what is the precise relationship between collective action by particular industries and the national interest? The posited concertation of approaches by both capital and labour certainly does not provide sufficient evidence of the public interest being taken into account.

Perhaps the most pregnant of Ronge's depictions of industrial politics is his equation of self-organization or self-regulation with a form of *Selbstverwaltung* (literally 'self-administration'). This is sufficiently important to justify a major digression on the nature and implications of German self-administration which can also be translated as 'self-government', more clearly indicating the political dimension. It will be recalled that the state-sponsored reorganization of the coal industry after the First World War was described as a form of *Selbstverwaltung*, and in the late 1920s the corpus of ideas systematically developed by Naphtali, Hilderding, and others, encapsulated in the notion of 'economic democracy' (*Wirtschaftsdemokratie*), also incorporated the fundamental concept of *Selbstverwaltung*. But this term has been borrowed from the sphere of politics, particularly from local government: *die kommunale Selbstverwaltung*.

Yet it was also demonstrated in Chapter 2 just how derivative *Selbstverwaltung* is as a concept. It implies a devolution or delegation of power and authority from central government, the heart of the state, and therefore from the primary institution of this German state which is the executive branch of government, the administration. The state as a bureaucratic power phenomenon, with its origin in monarchic authority, has been *the* institution which governed the evolution of German politics until at least the First World War, if not the Second. In German politics the dominant institution has not traditionally been parliament and its politicians but the state machine. Centralized, bureaucratic control, and democratic control and accountability were seen very much in administrative, organizational terms, with a commensurate influence, not least by example, upon the structure and organization of economic activity.[6]

German *Selbstverwaltung* in the Nineteenth Century

In the eighteenth century Great Britain offered the outstanding model of self-government, with democratic, representative forms of political association apparently characterizing every level of government—this contrasted starkly with the continental and in particular with the (Prussian) German system.[7] A thoroughgoing parliamentarianism was supposed to characterize the Anglo-Saxon system of government, monarchical bureaucracy the continental. In Germany, *Selbstverwaltung* was to become characterized by two principal elements: an essentially devolved, derivative quality; and bureaucracy in form and substance. The centralized executive branch of government was the dominant feature of the German state. German liberals, unlike their Anglo-Saxon counterparts, attempted to secure democratic rights and freedoms through developing the power of the judicature rather than the legislature: their bourgeois constitutionalism was more concerned with legal safeguards to control government than with effective parliamentary sanction.

These developments cannot, however, be divorced from the rise of Prussia, a state pursuing ambitious power politics, with its system of government determined by the demands of military organization. An exaggerated militarism and absolutism were manifest in their purest form, suppressing normal civic life. Frederick the Great established the indissoluble association of the monarchy with the East Elbian *Junkers*, consciously making them its main pillar of support. This military nobility (*Militäradel*) presided over the destruction of municipal autonomy and in the process municipal affairs became both bureaucratized and militarized. The French Revolution was not directly to exert a dynamic influence on the theory and practice of German government; the precedents established by a centralized, monarchical bureaucracy were too strong, even for France itself. Indeed, German *Selbstverwaltung* could not be considered from any perspective as possessing a revolutionary aspect; on the contrary it was characterized by timidity and conservatism.

A greater influence was so-called 'enlightened absolutism' which governed the climate of opinion at the end of the eighteenth century. In fact, after the death of Frederick the Great it gave birth to a so-called 'civil servant liberalism' (*Beamtenliberalismus*) which became a central factor in German politics as, especially from 1807 onwards, the upper echelons of the state bureaucracy were to acquire an increasing profile and independence.

However, the term 'enlightenment' represents a relative notion and Prussian enlightenment proved to be a double-edged sword as it expanded its influence during and after the Napoleonic wars. It was Freiherr von Stein who came to personify the strange, autocratic, authoritarian progressiveness of the time. Obstacles to reform were removed by the catastrophic defeats of 1806–7 and in 1808 he laid the foundation of all modern German *Selbstverwaltung* with his municipal order (*Städteordnung*): 'all further development is built upon it.'[8] Stein's innovations had been facilitated by the spread of the 'national idea' in response to the French invasion and the belief that greater unity was necessary in the organization of German affairs.

The thirty-three-year period between 1815 and 1848 was a period of restoration and reaction, the hey-day of Prussian bureaucracy. Prussia had become a 'civil servant state' (*Beamtenstaat*), one in which the alliance of bureaucratic monarchy and the feudal, militaristic *Junker* caste appeared most unshakeable. The whole system of bureaucratic rule, including especially the administrative élite, became more conservative, indeed, the most senior officials came to exercise a form of despotism.

Yet senior officials were to claim to be the natural mediators between the Crown and the people; the spokesmen of popular demands, they even felt that they embodied the idea of the state itself and, despite the monarchy, considered themselves to be the real governing class. Indirect support for this view was provided by the influence of Hegel's thought which to some appeared to glorify the Prussian state, particularly at the expense of the rights of the individual. Whatever the precise nature of Hegel's teaching, there can be little doubt that German Liberals generally were to display in the nineteenth century a paradoxical and most un-Liberal rejection of individualism. Yet at the same time as official intellectual opinion seemed to consolidate and buttress the power of central state bureaucracy developments were afoot economically and socially which were initially to undermine that very bureaucratic centralization. These took the form of industrialization and urbanization. Indeed, in the larger towns demands for local administration were such that the delegation of centralized power became unavoidable. It was precisely in these newer, growing urban communities that a new industrial and commercial class was to emerge with a self-confidence to demand political reform at local level and beyond.

The new industrialists held that purely local markets and regulations were inhibiting their business interests; a truly national market seemed a self-evident need, entailing also the political demand for national unity. But at every turn public officials seemed to be inhibiting expansion, forcing businessmen into the political arena to advance their interests. So political activity was forced upon the bourgeoisie and it is not surprising that bourgeois liberalism proved to be 'very moderate and decidedly undemocratic'.[9]

It was not until the 1840s that *Selbstverwaltung* became the widespread and accepted expression for local autonomy and lay involvement in local political affairs. By 1848, Liberals had both rejected thoroughgoing parliamentarianism and failed to establish a viable *Rechtsstaat* through strong, independent courts. The bureaucratic enemy had proved as resilient as ever, and 'precisely in Germany it was moderate Liberals who long after Stein remained in ignorance or principled rejection of English parliamentarianism'.[10]

Bourgeois attempts at governmental reform in the immediate aftermath of the 1848 revolution were guided by the intention to erect a system of thriving self-government at district, county, and provincial level, crowned by a national parliament. But prospects for reform were not favourable, with the *Junkers* organizing a new political party of their own (and establishing a secret shadow government), and with an authoritarian spirit living on in the army and state bureaucracy. The first leading politician to emerge from the new capitalist class was David Hansemann, who became a major public figure after the July revolution. In 1848, he actually became a Prussian Minister and attempted to introduce a comprehensive administrative reform. But it is very significant that this Liberal spokesman's main objective was bureaucratic reform, not a transfer of power and responsibility to elected, representative bodies: 'self-government' was 'self-administration'.[11]

The 1850s were to witness the eclipse of Liberal hopes in the political sphere as *Junker* feudalism proved stronger than at any time before or since. The main sources of Prussian reaction were the Crown itself, the landed aristocracy, the Conservative Party, the orthodox Protestant Church, and, above all, the class-conscious officer corps of the standing army. Feudalism and bureaucracy combined to repress Liberal *Selbstverwaltung*, restricting it within narrow limits. The new local government edicts of 1853 and 1856 contained two main objectives: extended central state supervision of local government, and an increase in the power of administration in relation to elected representatives at local level. These laws provided the legal framework and basis of German local government until well into the twentieth century, particularly in urban communities where the most spectacular developments occurred during the course of industrialization. Moreover, such laws were deliberately administered in an unconstitutional spirit, forcing Liberals to realize that the actual letter of the law could mean very little when interpreted in a totally illiberal manner. It seems that German Liberals were victims of their own formalism.

Liberals abandoned attempts to acquire governmental power—to capture the state—through parliamentary means and largely restricted themselves to advocating strengthened legal safeguards over the executive. They came to accept the leading role of the Crown and its ministerial officials and even to idealize the state as the highest authority, above all

THE GERMAN TRADITION OF SELF-REGULATION 157

groups or individual interests. And, not least under Hegel's influence, they began turning towards a form of romantic, nationalist, power politics. In consequence, in domestic politics German Liberals came to treat a comprehensive administrative reform as their main objective and the realization of this intention was to give substance to the modern meaning of *Selbstverwaltung*.

Gneist was the seminal Liberal influence on *Selbstverwaltung* during the third quarter of the nineteenth century and for some time after. For him the term meant the direct involvement of lay elements in the actual workings of administration. Gneist's central idea was that true *Selbstverwaltung* is based on honorary office-holders from the upper classes, a genuine *Honoratioren-Standpunkt*. His philosophy also granted special prominence to a state interest, and held that full parliamentarianism negated the state. Gneist clearly felt that *Selbstverwaltung* would be extended primarily on the initiative of the monarchy and its government of senior officials, these civil servants forming the governing class. His objective was a social union of bourgeois élite and aristocracy, together forming a broader, dominant upper class, controlling political affairs. His doctrine required honorary officials appointed by the state, as no rights existed independently of the state. The effect of Gneist's teachings was a blurring of the distinction between central and local government (*Staats- und Selbstverwaltung*), with the latter no longer conceived as a vigorous alternative but as a subordinate complement.

In 1858, William I assumed the Regency and it seemed in this new era that the period of reaction would soon pass. Liberals made certain advances in the field of administrative reform, the one area where Conservatives showed a willingness to compromise, but there was no compromise over the feudal privileges of the *Junkers*. The year 1866 marked the turning-point, with the Prussian victory over Austria at Olmütz. German Liberals were forced to recognize the supremacy of Bismarck and the style and direction of government in his hands. His victory, in both senses, guaranteed monarchic pre-eminence in government for the foreseeable future, a system based on the entrenched loyalty of the state bureaucracy and armed forces. This major set-back for the Liberals entailed the postponement of a fuller parliamentarianism for the foreseeable future, the very moderate nature of German Liberalism becoming reinforced, with more restricted notions of *Selbstverwaltung* serving as a major substitute in Liberal preoccupations.

On 13 December 1872, under the Second Empire, a new framework for county government in Prussian rural areas appeared on the statute book, a new *Kreisordnung*. Yet it would seem that the issue of *Selbstverwaltung* was to become overshadowed by the *Kulturkampf* and the social question in the ensuing decades. However, industrial expansion, population growth, and the concomitant development of interests and their expression in associations and organizations merely ensured that self-regulation would acquire a new dimension. Catholicism and its political representation, the Centre Party, were bound to become of crucial importance, not least as the focal point of resistance to Prussianism and the Bismarckian system. The Centre Party represented a much stronger opposition to Prussianism than moderate Liberalism, and German Catholicism had developed a distinct self-government tradition of its own. So in matters of constitutional law the Liberals and Catholics were *de facto* allies, despite their religious differences. In addition, the labour movement's trade unions and consumer co-operatives also represented the Liberal idea of co-operative self-help.

But just as new forms of *Selbstverwaltung* were emerging to counteract Liberalism's waning force, the economic crisis of 1873 and its aftermath stimulated the tendency nurtured by absolutism to appeal to state authority. Bismarck's opting for trade protectionism could be interpreted as a return to a form of Frederician mercantilism. His measures of social insurance were also to be supported by the Socialists of the Chair (*Kathedersozialisten*, mostly right-wing Liberals) mobilized through the Society for Social Policy, who believed that social reform would draw the teeth of social democrat-led class warfare and neuter the threat to the existing order. Their antipathy to representative, parliamentary democracy, like that of Gneist and Treitschke and personified in the figure of Gustav Schmoller, lent strong indirect support to Bismarck. Again, the Liberal bourgeoisie was more afraid of a popular franchise and the labouring classes than of aristocratic authoritarianism.

In terms of meaningful devolution the Provincial Order of June 1875 represented a disappointment and marked the end of hopes for a wide-ranging provincial decentralization. It confirmed the nature and implication of the earlier County Order: domination by conservative bureaucracy was as pronounced as ever and the clear Anglo-Saxon distinction between civil servant/public official and parliamentarian/councillor, based on a relatively unambiguous distinction between the executive and legislative branches of government, remained as blurred as ever in the practice of German politics. But *de facto* dominance by the *Junkers*, steadfastly resisting nearly every reform, was not entirely negative: whatever the social composition of Germany's civil servants and public officials, the dominance of the executive over the legislature and the ready interchange between the two facilitated, in a technical sense, efficient government, not least through the minimizing of partisan politicking.

The defeat of the Prussian Liberal initiative of 1876 to modernize urban local government was significant in two ways: it marked a major defeat for German Liberalism at the very outset of the new Reich, and it meant that by far the most important area of local government[12] was to be left unreformed (von Stein's system, with the modifications of 1853, remained the basic framework). The exclusion of built-up areas from local government reform was all the more remarkable as Germany's economic expansion was concentrated primarily in towns, with the accompanying population growth creating an urban explosion. As the state took over the post office and the railways, local authorities were establishing or taking over gas- and waterworks, trams, slaughter houses, and so on—a municipal socialism was appearing alongside a state socialism well before this was advocated by the radical reformers of the Fabian Society in England. Local authorities were compelled by events to expand and develop their own professional staff, a reflection of a much wider trend towards the bureaucratization of public life, a general consequence of the industrial revolution. The German labour movement was by no means excluded from this development, proceeding to develop a formidable bureaucratic apparatus of its own.

In contrast to state and national levels where bureaucrats dominated elected representatives, municipal administrators were elected by the local council and were obliged to co-operate with it. For this reason municipal bureaucracy was noted for its Liberal (mostly patrician) character. Yet central government had no inhibitions in ruthlessly applying its right to authorize and disallow nominations for the senior posts in local administration.[13] Although the Progressive Party dominated in many urban communities in Prussia, only in exceptional cases was this reflected in the appointment of left-wing Liberal nominations to the most senior administrative posts. Just as at national level the Free Conservatives were the most favoured party as a source for top bureaucrats, so at local level it was the right-wing Liberals, the National Liberals, who dominated. Indeed, the prevalence of National Liberals holding senior administrative posts in local government, enjoying substantial salaries and prestige, was a major factor in undermining demands for a genuine parliamentarianism.

Although Liberal success in extending meaningful *Selbstverwaltung* was extremely limited, progress in legal reform yielded more positive results. Improvements in administrative law (*Verwaltungsgerichtsbarkeit*), in a direction long since paved by France, were complemented at national level by standardized reforms of the general legal system (*Justizreform*). Together, they represented one of the few major achievements of nineteenth-century German Liberalism.[14] However, such reform as had occurred only served to remove the worst anomalies and abuses, thereby entrenching the existing system, particularly in the heartland of Prussia— by renouncing full parliamentarianism German Liberals had themselves guaranteed that *Selbstverwaltung* would remain a subordinate force.

A Century of *Selbstverwaltung* Reviewed

By the end of the nineteenth century German classical Liberalism had been eclipsed as a political philosophy and Bismarck had destroyed the once great National Liberal Party. The conflict of principle between state sovereignty and popular sovereignty had long been decided in favour of the state. Gneist's doctrine that true *Selbstverwaltung* consisted in incorporating the lay element into government administration and giving an equal status to honorary positions had been rendered nugatory by events. By the early twentieth century the generally accepted view of *Selbstverwaltung* was of 'independent administrative activity by local authorities in the sphere of their own interests and responsibilities, yet integrated with the state, especially in being subject to its supervision'.[15] This apparently final consensus on the real nature of *Selbstverwaltung* seemed to ignore other forms of *Selbstverwaltung* outside local government, such as working-class self-government. Excluded from political power, the working class developed a very active form of self-organization in its own trade unions and co-operative societies. In fact, the German labour movement was beginning to assume the mantle of the independent spirit of self-government. The SPD, as the party of the urban proletariat, was to assume the vanguard role regarding constitutional reform relevant to a mass industrial society. It was social democracy, in alliance with the left-wing Liberals, which was to carry *Selbstverwaltung* to its logical conclusion after the First World War: 'The dualism of state "administration" and self-"administration" was only to be overcome in the genuinely free state of parliamentary democracy'.[16]

Outside the sphere of local government, *Selbstverwaltung*, as a form of apparently fairly autonomous self-regulation, was gaining strength in the commercial world in the employers' associations, the great industrial cartels, and especially the RWCS.[17] The very first expression of modern self-regulation within the German economy had been inaugurated, like so many major innovations in political and economic spheres, by the French occupiers, in the Chambers of Commerce. They were retained by the Prussian authorities and during the *Vormärz* period were extended to the rest of the state, eventually establishing an elaborate network of Chambers of Commerce, Industry, Agriculture, and Handicrafts. In 1843 Bavaria followed suit.[18]

Between the state sphere of government and the non-state sphere of industry and labour a major, additional semi-state form of *Selbstverwaltung* had been created in the 1880s, in the sphere of social insurance. Hitherto, under the influence of classical liberalism, the principle of self-help had dominated official thinking in this area. Now the state was to transcend that principle with a form of *Selbstverwaltung* approximating to that of local government. The state was largely breaking new ground although, among the few precedents which existed, there were the friendly societies of the miners and steelworkers.

The first measure was the Act of 15 June 1883 regarding general sickness insurance based primarily upon independent occupational schemes. The new organizations were not subject to rigorous state supervision and, as the workers provided through their contributions three-quarters of the income, they became the dominant influence. This influence was exerted in practice by the social democrats, who experienced their first major participation in public administration. The measure was followed by the Act of 6 July 1884 regarding accident insurance in which the occupational schemes of particular branches of industry were extended to cover the whole Reich. As the employers met all the financing themselves, they alone provided the personnel, and workers' influence was extremely limited. This particular instance of self-regulation was subject to the supervision of the Imperial Insurance Office, a bureaucratic institution soon to develop a wide-ranging and detailed involvement in the expanding social insurance system.

The Act of 22 June 1989, covering pensions for old age and disability had no precursors and greatly extended the number of insured; it included, for example, handicraft journeymen, agricultural labourers, and domestic servants from its very inception. Finance was provided equally from three sources: the state (Reich), the employers, and the workers. The scheme was organized on a state-wide basis and, as with the accident insurance scheme, there was 'a strong bureaucratic element beside the element of corporative [*genossenschaftlich*] self-government'.[19]

Later regulations, such as the Imperial Insurance Order of 1911 and the White-Collar Employees' Insurance Scheme (*Angestelltenversicherung*), also of 1911, confirmed the Bismarckian trend of social insurance organization which became more bureaucratic and more centralized. The element of *Selbstverwaltung* had become secondary well before the outbreak of the First World War. The later schemes possessed not only the name but also the character of an institution (*Anstalt*), and the main administration was in the hands of professional civil servants (*staatliche Berufsbeamten*); the lay element was restricted more and more to an advisory and supervisory role. So the organization of social insurance itself appears after all not to have given a special impetus to corporative self-regulation within the industrial economy and society.

Industrial *Selbstverwaltung*

The Weimar Constitution created a general politico-legal context for genuine, representative parliamentarianism and in particular the abolition of restricted franchises marked the end of a neutered mass democracy. The issue of *Selbstverwaltung* ceased to be a political phenomenon of major significance once the principle of state sovereignty had finally been replaced by the principle of sovereignty of the people. But the rise of the German labour movement to political prominence ensured that the issue of self-government would acquire a new dimension, in the economic sphere.[20] The revolutionary political climate in the immediate post-war years exerted pressure on national politicians so to order economic life that, in general, democratic principles would acquire relevance in the management of industry and that, in particular, organized labour would attain a recognized and meaningful position in sharing in determining the direction of industry.

The more radical elements within German social democracy were demanding a thoroughgoing socialization of key industries; in particular, the movement for soldiers' and workers' councils demanded expropriation and the introduction of a grass-roots, participatory form of management for socialized industries. These demands represented part of a comprehensive, if uncoordinated, political, social, and economic programme of German national reconstruction. Yet largely through the efforts of one man with influence and support in certain left-wing circles, the movement in question was diverted largely into a preoccupation with purely economic matters at the expense of political ones. That man was Richard von Möllendorf, formerly Deputy State Secretary at the Imperial Economics Ministry under August Müller and Rudolf Wissell, the latter becoming (SPD) Economics Minister in 1919. In the immediate post-war period it was Möllendorf who proved to be the principal architect of plans and legislation regarding socialization and *Selbstverwaltung*. His influence was such that it is analysed extensively in the following pages.

Möllendorf was a *Junker* of impeccable social lineage and had been called to head the Imperial Economic Office at the end of the war. This office, a department of the Imperial Economics Ministry, was primarily responsible for overseeing the transitional arrangements from a wartime to a peacetime economy. It was Möllendorf who created the term 'commonweal economy' (*Gemeinwirtschaft*), who consciously applied its principles to legislation, and who drafted the Socialization Act. Möllendorf was no passive administrator, executing the instructions of his political masters, but an active initiator of a new economic order.

Möllendorf treated the commonweal economy and industrial *Selbstverwaltung* (self-government) as inseparable and, above all, interpreted the new form of industrial organization as an intermediate form between state ownership and private ownership. Such a system would more readily take account of national imperatives and the form of implementation it would take would be that of *Selbstverwaltung*. Central to this form of industrial organization is the attempted reconciliation of a public or national interest with individual, private interests, and representative bodies to realize this synthesis.

As has already become evident, the wartime experience had a seminal influence in shaping the attitudes of government and employers towards industrial organization, in particular in enhancing the sense of responsibility of private employers towards the so-called national interest. It is quite clear that the state initially bypassed private employers, attempting to direct the economy without employer assistance; it was for technical reasons, when this system patently failed to work, that private industrialists became intimately involved in economic policy decisions. It was not indignant employers deliberately boycotting the system which brought about necessary changes but sheer practical necessity when civil servants and the military demonstrated their inability to run industries. Equally important was the strengthening of the tendency which had already existed before the war for industrialists to attempt to settle general industrial problems on a joint, communal basis.[21] German professional organizations, although clearly conceived initially as interest groups, during the course of the war exercised the important public function of restricting the conflict potential of domestic competition, and employers' organizations became used to carrying out explicitly public functions.

No sooner had war broken out than it became almost the daily practice of government to consult trade unions, as the official representation of labour, although no constitutional basis for this existed. Government also utilized the employers' associations, creating such organizations where none had existed before, for example in footwear and soap. But it was the raw materials sector which was to prove decisive. The principles of organization of the commonweal economy were first developed to manage Germany's scarce raw materials and to undertake any necessary reorganization. The practices thus established gradually spread to other sectors of the economy and were continued after the war in the coal industry (Germany's major natural resource). The commonweal economy was exclusively the creation of the recent war and its aftermath.

Commonweal economy was not a legal concept but an economic one and one which was opposed to economic individualism. There was no Smithian belief in an invisible hand, guiding individual economic actions so that their total sum coincides with the best national interest. The optimal allocation of resources was a moral matter as much as a technical one, and in the process of decision-making the state was to play a central role.

But how? Certainly not in the traditional manner. The degree of state control accompanying state ownership and the new wartime controls over private industry were considered too extreme and therefore inappropriate, so 'this special kind of commonweal economy had arisen in contrast to the state economy'.[22] And, as seen above, state bureaucrats had been forced to recognize their own limitations and, increasingly, to involve industrialists in economic decision-making, with the result that occasional participation gradually led to enclosed industrial groupings (*geschlossene Wirtschaftsgruppen*) assuming key responsibilities and the state granting them public authority to carry out the necessary measures.[23] Legislation in the immediate post-war period attempted to continue and develop this idea and form of organization. In the administrative idiom of the local government tradition: 'The commonweal economy should not be "state/government administration" [*Staatsverwaltung*] but be conducted as a form of self-government/"administration" by self-governing bodies'.[24] Although large organizations, by definition, have to be administered, it was not considered that the norms and practices of the executive branch of central government were transferable to the organization of productive industry.

The principle of solidarity, both as a means and an end within the new form of industrial organization, was by no means without precedent, as Möllendorf was fully aware; yet not even syndicalistic workers' enterprises, any more than the great cartels, guaranteed that, in the final analysis, selfish interests would not take precedence over public ones. Hence the key issue became one of precisely how and by whom the economy was to be regulated; it was undoubtedly to be regulated. The wartime precedents were ineradicable, with the economy as a whole having become organized as a public economy, with no substantial changes in ownership.[25] The state as representative of the national interest took control of the private sector and, as we have seen, it was also considered essential that the general (national) interest should remain dominant in the immediate post-war years. The commonweal economy had come to mean the following: 'Commonweal economy . . . includes any economic organization regulated by the public power.'[26] This could extend to full socialization, which was not the same as nationalization, where an industry became part of the state machine, like the post office and the railways. The wider interest was generally to be secured by state supervision and by the presence of consumers on the appropriate cartel bodies. It was also generally assumed that labour would receive parity representation there, too, but this was not inherent in the concept.[27]

Soon after the First World War the bodies of economic self-government had come to assume the form of parliament-like organs consisting of 'representatives of an industrial grouping . . . mostly producers, dealers, and consumers . . . , and observing complete numerical parity between employers and employees';[28] again, this followed the wartime precedent. The local government model was quite explicit:[29]

> regarding the economic groupings . . . it is a matter of corporative associations of a public-law nature integrated with the state and which exercise functions independently of the state or let them be exercised by their representative bodies and which are in the possession of public-law rights and duties . . . like the exercise of public powers in another sphere by free, corporative associations, integrated with the state, liked the local authorities and local governments associations . . . only with the difference that for the latter the corporations are constructed upon a personnel and territorial basis, but for the former an economic basis had replaced the territorial.[30]

Years of evolution in local government practice, and not least the influence of the legal thinker Rosin, had contributed towards greater clarity in perceiving the precise nature of *Selbstverwaltung* and its umbilical relationship with the state. The problem of the precise nature and form of state involvement was crucial. The wartime trend increasingly to involve industrialists and simultaneously to pass public responsibility to private industrial organizations came to be welcomed by the industrialists involved. They added their voices to a demand for more *Selbstverwaltung*, such a call being perfectly compatible with Möllendorf's plans: 'The whole of state economic policy was to be exercised predominantly by industry itself.'[31] In fact, his ultimate objective was ' "the abdication" of the state in this sphere in favour of industry regulating itself, merely with the proviso of a right of supervision for the state'.[32] Such an intention was compatible with the wishes of major industrialists who around 1920 were particularly vocal in their own demands for an organic economy, a commonweal economy, and industrial self-government.

However, for Möllendorf, state supervision was not intended as an empty formality, as ultimate authority was to lie with 'a state to whom sovereignty would remain in all economic matters'.[33] After all, just as with local authorities, organs of industrial self-regulation owe their legal status to state action. A commonweal economy was to be continued after the war not just by the creation of appropriate organs of industrial self-regulation for particular industries, but also by the creation of an Imperial Economic Council and of District Economic Councils grouped into areas or provinces.

As is now an accepted consequence, the wartime experience proved seminal. The Imperial Economic Council was regarded as 'the ultimate consequence of the idea of a public economic leadership, of a responsible participation in state economic policy by the industrial circles immediately concerned (in this respect also industrial self-government)'.[34] Similarly, the new organization was to relieve the state in a manner parallel to that which had been developed by the local authorities, the economic councils to become responsible for the basic principles and guidelines of public policy. But whatever the intention of its protagonists the Imperial Economic Council actually created was not to fulfil its potential although, according to Glum, it possessed in reality a stronger position than the Constitution appeared to give it.

Private industry was prepared to tolerate certain ideas, and one or two organizations, associated with the philosophy and principles of the commonweal economy during a period of acute political and economic uncertainty, as this was an important tactic for minimizing hostility and facilitating survival. But the co-operation of industrialists began to wane as the political position of social democracy weakened sharply at the very beginning of the 1920s and, with economic consolidation from the mid-1920s onwards, the commonweal economy, with its publicly sponsored organs of self-regulation, faded into obscurity. Glum had confidently predicted in 1925, when his work appeared, that the Economic Councils would not survive in any major way. After all, employers had never been very keen on the limitations which the commonweal economy entailed, and dedicated socialists had wanted a much more thoroughgoing socialization; neither side of industry had been even remotely satisfied. By the mid-1920s it had become unmistakable that 'The private sector had proved to be stronger than the commonweal economy.'[35]

Yet although a commonweal economy was not established, the consensus by 1924/5 being 'that the realization of the idea of economic individualism [*Individualwirtschaft*] furthered the national economy more',[36] the public responsibilities of industrialists came to be generally accepted, along with an appropriate organizational form for their concrete expression.[37] The German economy was to be highly organized, but by whom and in whose interest remained unclear. Möllendorf's objective had been that of 'an economy organized through and through' ('einer vollständing durchorgnisierten Wirtschaft'). Chapter 2 of Part I has described the form which the economy was actually to assume. Although German industry did evince a considerable degree of autonomy and self-organization in the later 1920s, a minimal role for the state had been inconceivable in principle, for Glum referred to the German view of the state (admittedly with some recent revision) as being one 'in which we are accustomed to seeing the model for all organizations'.[38] The state was not purely a political institution, narrowly circumscribed by law; the moral nature of the state as an institution justified its ultimate authority within the wider society.

It was held, regarding self-regulation within society as a whole and not just in the economy, that: 'It will be the task of politics . . . to prevent the sovereignty of the state from being impaired by these new organizations, so that their activity remains within the framework of self-regulation.'[39] Nevertheless, the precise boundaries were ambiguous and the concept of *Selbstverwaltung* was sufficiently flexible that all associations could retain a large measure of freedom, especially regarding their legal structure. Apparently, what was valid for local government was also valid for the economy: 'The idea that "the free community" was the foundation of a free state should also be the directing principle for the relationship between the economy and the state.'[40]

A New Post-War Consensus?

As has been demonstrated, the later 1920s witnessed a capitalist reconstruction and resurgence. The organization of industry was undertaken largely according to private initiatives, self-regulation coming to incorporate a large measure of industrial autonomy, with the formation of various kinds of monopolies, for example cartels and trusts. But the economic collapse of the Great Slump shattered business confidence and marked the re-entry of the state into a more dominant position within the economy. The early 1930s represented a mirror image of the early 1920s. Just as the 1920s marked a period of transition between wartime state regulation and restored entrepreneurial power, so the 1930s came to represent the transition back to wide-ranging state control, consolidated under German Fascism by the Nazi dictatorship.

In retrospect, one can see how the evolution of the German political economy during the inter-war years represented continuing variation on a constant theme, that of an organized capitalism:

'Organized capitalism' is characterized by a 'concentrated' and bureaucratic, association-led system of economic organization whose functioning is secured by state measures of various kinds and which takes the place of a competitive economy sustained by individual firms and largely autonomous regarding state intervention.[41]

The concept and practice of *Selbstverwaltung* was sufficiently flexible to incorporate substantially varying degrees of relative state influence within the overall system of organized capitalism. This was not the case, however, with economic democracy. Glum had maintained that the form of self-regulation entailed by the commonweal economy would ultimately be characterized by the abdication of the state, that a self-regulated economy would be merely subject to state supervision, the realization of which would also be called economic democracy. But economic democracy had come to possess the representative connotations of political democracy involving consultation of, and dependence upon, the wishes and decisions of all participants stretching from citizens in the polity to employees (including managerial staff) in industry. This had certainly not been welcome to the employers, especially as the primary impetus for economic democracy was emanating from German socialists, i.e. the social democrats, whether as members of a particular party or as trade unionists.

It is clearly no coincidence that one of Naphtali's co-authors in his major redefinition of economic democracy was no less a politician than Rudolph Hilferding (SPD), the originator of the concept of organized capitalism. Notions of the commonweal economy and of organized capitalism were certainly compatible with one another, and so various forms of self-regulation could also be incorporated. But the evolution of self-regulated, organized capitalism between the wars did not extend to economic democracy. It appeared largely to ignore workers' interests and institutions. For Möllendorf, involving labour, apart from appeasing the workers' councils movement, had been largely a pragmatic tactic to engage all forces which were lying fallow in the national task. For socialists, in contrast, worker participation was a moral, democratic imperative as well as practically essential in improving production.

The period following the Second World War saw an extended flirtation with a socialist vision of a commonweal system before reverting to the traditional norm. Few post-war critics of the adverse consequences of the pre-war system were more acerbic than Konrad Adenauer of the CDU:

we certainly take the view that Ruhr industry ... in the years before 1933 politically exploited the great economic power which becomes concentrated there to the detriment of the German people.[42]

He characterized organized capitalism as 'the entwining [*Verflechtung*] of the economy [which] is ... the cause of much disaster which has befallen us'.[43] The Liberal Finance Minister (NRW) at that time, Blücher (FDP), also stated: 'We are not considering forgetting the sins which were perpetrated precisely in this sphere by monopoly capitalism and excessive greed for profit.'[44] With leading representatives of the so-called bourgeois parties thus inclined, one could expect that the general climate of opinion would be very susceptible to socialist ideas and proposals.

The aftermath of the Second World War witnessed far greater agreement among the political parties about the basic principles and immediate practical priorities regarding post-war reconstruction than had been the case after the First World War. There are few instances where the political consensus was more clearly demonstrated than in the way basic notions of commonweal economy appeared to have become the conventional wisdom in industrial politics. On only the second day of the very first session of the new North-Rhine Westphalian *Landtag* a unanimous resolution was accepted which contained the claim: 'The transfer of mining to a commonweal system of organization under German responsibility and management is necessary.'[45] It was signed by Adenauer on behalf of the CDU, Henssler (SPD), Middlehauve (FDP), Reismann (Centre Party), and by Kaiser on behalf of the KPD. Admittedly, the primary concern was to regain German control of the coal industry; nevertheless, that the new form of organization was to involve a commonweal economy solution was self-evident to all concerned.[46]

However, there can be little doubt that the main inspiration for commonweal economy remedies emanated from the SPD. Indeed, during the fifth session of the North-Rhine Westphalian *Landtag* (during the so-called *Ernennungsperiode*), the SPD officially advocated that, at the same time as the first elections for the *Landtag* took place on 20 April 1947, a referendum be held on transferring coal, iron and steel, and basic chemicals (*Großchemie*) into common ownership (*Gemeineigentum*). The common ownership of coal was also supported by the Centre Party. But how did the SPD conceive its objective of a genuine commonweal ecomomy for basic industries? An investigation of this question reveals nothing but ambiguity, contradiction, and confusion.

The first major SPD submission to the *Landtag* regarding coal socialization, submitted during July–August 1947, was referred to the House Economic Committee for further consideration. The SPD's motion was finally accepted by the Economic Committee on 31 March 1948. Thereupon the SPD submitted a further motion specifically related to closer specifications of the precise construction and functioning of the organization Selbstverwaltung Kohle. This led to further deliberations in the Economic Committee, especially as there appeared to be a great desire on all sides, but particularly within the SPD, to secure the greatest measure of unanimity. Not only was the Centre Party strongly in favour of the socialization of coal the CDU was also on record with an appropriate motion of its own.

The Economic Committee finally submitted the approved draft bill on coal socialization to the *Landtag* assembly on 22 July 1948. The law was finally passed (incorporating certain further amendments) by a large majority on 6 August 1948. However, after these fifteen months of countless discussions, motions, and debate the British Military Governor informed the *Landtag* of North-Rhine Westphalia that its resolution on the said bill was null and void, primarily on the grounds that coal was a national industry which could only be reorganized by a national German government, arising out of free national elections. It was a sudden and definitive *coup de grâce*.

Did this mean the end of hopes for a commonweal economy in general and a self-regulated or self-governing coal industry in particular? Not necessarily, the CDU was officially on record as claiming:

The capitalist economic system has not proved equal to the vital interests of the German people in state and society ... Through a commonweal economic system the German people will obtain an economic and social constitution corresponding to the rights and dignity of man, serving the moral and material reconstruction of our nation and securing peace internally and externally.[47]

However, if one considers the actual voting on the final coal bill, grounds for suspicion arise. Despite the long and arduous negotiations to secure maximum unity within the *Landtag*, the FDP voted against and the CDU abstained (with one CDU Member voting with the FDP). It was precisely this bourgeois alliance of Conservatives and Liberals which was to come to power in 1949 and to go from strength to strength during the 1950s. The Centre and Communist Parties, which had backed the SPD, were soon to become totally eclipsed, so clearly the writing was on the wall for the commonweal economy (but not necessarily for *Selbstverwaltung*, a very flexible and hardy perennial). It is almost as if the bourgeois politicians had been as similarly motivated as the captains of German industry after the First World War, deliberately adopting delaying tactics for as long as was necessary. As Kirdorf said on the fiftieth anniversary of the founding of the Rhenish-Westphalian Coal Syndicate: 'Yes, we went into the socialization committees in order to prevent socialization.'[48]

Notes

1. V. Ronge, ' "Solidarische" Selbstorganisation der Wirtschaft—eine Alternative zur Politisierung im Spätkapitalismus', *Leviathan*, 2 (1978), 183.
2. Ibid.
3. Ibid. 184.
4. See V. Ronge, *Am Staat Vorbei* (Frankfurt: Campus Verlag, 1980).
5. Ibid. 24.
6. See H. U. Wehler, 'Der Aufstieg des Organisierten Kapitalismus und Interventionsstaates in Deutschland', in H. Berding *et al.*, *Organisierter Kapitalismus: Voraussetzungen und Anfänge* (Göttingen: Vandenhoeck & Ruprecht, 1974), 46–7.
7. The present section is based closely on Heinrich Heffter's seminal work: H. Heffter. *Die deutsche Selbstverwaltung im 19. Jahrhundert* (Stuttgart: K. F. Koehler Verlag, 2nd edn., 1969).
8. Ibid. 11.
9. Ibid. 230; see also ibid. 91.
10. Ibid. 86.
11. The most historic innovation for Prussia at this time was the introduction of the Three-Class Voting system (of 30 May 1849), an 'undemocratic' franchise applying to all local, regional (provincial), and (Prussian) state elections.
12. At a conservative estimate, its expendure was at least ten times that of the counties.

13. As early as the 1870s some social democrat councils had been elected in certain industrial parts of Saxony, but the government refused to accept their nominee for mayor.
14. However, police power (*Polizeigewalt*) remained in the hands of the Prussian government, exercised under the unlimited powers based on laws enacted by Frederick the Great.
15. Heffter, *Die deutsche Selbstverwaltung*, 707.
16. Ibid. 182. For the role and influence of Hugo Preuß, see ibid. 753–4.
17. But although the associations in question were largely private ones, the autonomy of the newly emerging employers' associations was not quite as complete as might be expected: despite being self-governing bodies, as public-law organizations they nevertheless became integrated with the state (ibid. 271).
18. These institutions, seen historically, provided an intermediate, organizational link between the guilds of pre-industrial times and the aforementioned employers' associations and monopoly systems.
19. Heffter, *Die deutsche Selbstverwaltung*, 690.
20. This section is based closely on F. Glum, *Selbstverwaltung der Wirtschaft* (Berlin: Hermann Sack Verlag, 1925).
21. '... in Germany, in every sector of the economy ... a finely constructed network of interest groups and specialized associations developed, to an extent that can hardly be matched anywhere in the world' (ibid. 44; also 57).
22. Ibid. 18.
23. In this way: 'A special form of commonweal economy had developed out of the state economy' (ibid.).
24. Ibid.
25. Ibid. 14.
26. Ibid. 17.
27. Article 156, para. 2, of the new Constitution, though, expressly envisaged organizations of this kind which were soon to be created for coal and iron and steel.
28. Glum, *Selbstverwaltung der Wirtschaft*, 21.
29. The standard legal definition of *kommunale Selbstverwalrung* was by now well established: 'exercise of public-law powers by public-law associations which are integrated with, and subordinate to, the state, but which are, notwithstanding, independent of it' (ibid. 28).
30. Ibid. 27–8.
31. Ibid. 55.
32. Ibid. 24.
33. Ibid. 54.
34. Ibid. 130–1.
35. Ibid. 12.
36. Ibid. 64.

THE GERMAN TRADITION OF SELF-REGULATION 173

37. Ibid.
38. Ibid. 176.
39. Ibid. 177.
40. Ibid. 178. But this freedom was conceived very much within the German tradition, one dominated by the 'idea that the political strength of the German state was based on a corporative [*genossenschaftliche*] substructure and on a healthy mixture of freedom and compulsion' (ibid. 50).
41. Dust-jacket blurb from H. Berding *et al.*, *Organisierter Kapitalismus* (Göttingen: Vandenhoeck & Ruprecht, 1974).
42. Official minutes of the North-Rhine Westphalia *Landtag* debates (*Ernennungsperiode*). Official archive copies, 23 Jan. 1947, 10.
43. Ibid. 4–6 Mar. 1947, 20; along with 'the concentration of economic power in the hands of few people'.
44. Ibid. 23 Jan. 1947, 8.
45. Ibid. 13 Nov. 1946, 51.
46. For months later Adenauer demanded such a solution for all the main basic industries, the commanding heights of the economy which seemed to form natural monopolies (ibid. 4–6 Mar. 1947, 21).
47. Transcript of Motions submitted to the *Landtag*, Ref. I–109, 35.
48. Quoted by a Communist speaker; see the Official minutes (1. Wahlperiode), 6 Aug. 1948, 968.

8

Coal and Self-Government

With the emergence of an affluent society and the establishment of relative political stability under a bourgeois political alliance during the 1950s, any sense of urgency regarding the necessity for a commonweal solution to the problem presented by the coal industry tended to disappear. The main overwhelming priority seemed to be largely a technical one, namely maximizing output to meet the national and European shortage of energy. It could strike a superficial observer that very little had changed in the German coal industry compared with the pre-war years. After all, the creation of influential coal employers' associations had been allowed; the re-establishment of extensive coal–steel integration had been permitted; and, above all, powerful cartel-like selling agencies had been conceded. However, two fail-safe mechanisms had been created which, taken together, would seem to guarantee that the German *montan*-complex would never again assume a form which could threaten political democracy, namely, the introduction of parity co-determination on the boards of *montan*-companies (plus a labour nominee as social director on the board(s) of management), and the limited internationalization of the *montan*-complex through the establishment of the European Coal and Steel Community.

The British occupation power, with a Labour government, had taken a view, identical to that of German trade unionists in general and of social democrats in particular, that a substantial presence of labour representatives in the highest decision-making bodies would prevent the emergence of reactionary, authoritarian industrial structures facilitating a renaissance of totalitarianism. The intention found general expression in the words of Dr Nölting (SPD, North-Rhine Westphalian Economics Minister):

> More important and essential than the question of ownership is how working people are to be integrated into the management of production . . . Only the new directing bodies in which employees are incorporated, give socialization its socialist content.[1]

It found specific expression in a resolution passed at a mineworkers' conference at the end of 1948. Noting Law No. 75 and criticizing the fact that the DKBL's directorate would not have equal employee representation

COAL AND SELF-GOVERNMENT 175

(unlike its Coal Advisory Council (*Kohlenbeirat*)), the resolution empowered the union's executive committee:

[to] present to the Military Government the urgent demand of the delegates for equal engagement [*Einschaltung*], of mineworkers in every organization of the mining sector in the proposed reorganization of mining.[2]

Whatever the significance and relative merits of parity co-determination, the supranational controls established by the ECSC were intended to play a part in preventing Ruhr heavy industry from again becoming a basis of aggressive nationalism. But once the energy shortage had been transformed into a surplus, within the context of reasonable political stability in the Federal Republic, and with the *montan*-sector about to lose gradually a degree of its strategic importance within in the total economy, the ECSC was to prove virtually an irrelevant organization. Saving the German coal industry, with tacit international approval, reverted to being essentially a domestic, national problem.

Attempts to resolve the German coal industry's ensuing crises would, however, require an appropriate framework of organizations. Existing employers' organizations, such as those existing for purposes of wage negotiations and to promote and organize research, were clearly inadequate and new organizations had to be created. For clarity of exposition, the present chapter distinguishes between employer and labour (essentially trade union) approaches. Co-ordinated management of the endemic crisis only became truly possible with the creation of a unitary company for the Ruhr, the Ruhrkohle AG, with the attendant merging of employer and labour perspectives and organizations. This merger of interests was related to the institution of co-determination introduced in 1951 and in general these developments did not bypass the state. The sense in which they represented an alternative to state action has already received critical attention.

The Non-Labour Dimension

The German Mining Emergency Association

The unity of purpose which had characterized the German coal industry for so long had emphatically re-emerged during the 1950s, its tradition of joint, common action proving to be as resilient as ever. The industry's own form of *Selbstverwaltung* was presented with a new and ominous challenge with the onset of the first coal crisis. The first organized, self-regulatory response was the creation of the German Coal Mining Emergency Association Co. (Notgemeinschaft Deutscher Kohlenbergbau GmbH). This organization represented no less than forty-eight mining and iron and steel companies. Its management team of ten members was headed by Kurt Haver, managing director of the Ruhr Coal Advisory Agency Co. (Ruhrkohlen-Beratung GmbH) and included Dr Heinz Reintges of the Ruhr coal employers' association (UVR); its advisory council (*Beirat*) was chaired by Dr Hans-Werner von Dewall from the Board of Management of the Hibernia Co., who was also vice-chairman of the UVR.

Coal importers during the later years of the energy shortage had entered upon long-term contracts for coal imports from American suppliers, at prices above Ruhr coal prices; then freight rates dropped dramatically taking the cost of American coal well below the German prices. The coal importers hastened to sign new long-term contracts at the cheaper prices, not least to compensate for the dearer contracts, and this necessarily entailed a diminished market for domestic German coal. Since 60–70 per cent of the coal contracted was accounted for by importing agencies owned by German mining companies it was agreed to compensate the agencies for cancelling the contracts (which contained a $2 per ton default penalty) and to replace them with coal of the same quality at the same price. This deal was effected by the Notgemeinschaft. For the whole operation to have succeeded so smoothly and expeditiously a remarkable degree of inter-company cohesion was indispensable; that the necessary co-ordination was both possible and effective was guaranteed by the other responsibilities and experience of the personnel entrusted with the task. The backbone of the staff was provided by the Ruhrkohlenkontor, the joint office of the Ruhr coal sales associations, of which Haver was the chairman (he was also chairman of the Ruhrkohle Trust Company).

Whatever the capacity for efficient self-organization which existed here within the industry, it has been sufficiently demonstrated that German *Selbstverwaltung* does not operate in a state vacuum. So it is no surprise to hear Reintges's full tribute to this particular instance of cohesive state–industry interaction and co-operation; nor is it surprising to learn that the creation of the Notgemeinschaft was on government insistence in the first place.[3] The whole scheme could not possibly have functioned without the coal import tariff which raised the price of third-country coal above the domestic price, irrespective of prevailing international freight rates. The financing was covered by government-backed credit guarantees, an arrangement effected by a consortium of banks headed by the Rhenish Giro Centre, Düsseldorf. State–industry co-operation, at least in the coal

COAL AND SELF-GOVERNMENT 177

sector, was as effective under the aegis of a social market economy as perhaps at any time previously in German history, as later developments were to confirm.

The next joint organization, founded on 21 January 1960, the Ruhr Mining Action Co. Ltd. (Aktionsgemeinschaft Ruhrbergbau GmbH) was short lived because of objections raised by the High Authority. The intervention of the ECSC induced the Federal government to give its attention to a new scheme. It was still believed that the overall level of German coal production could be maintained provided rationalization was given adequate priority, a process which would only be possible with concentration on the most profitable and therefore apparently promising pits.

The Rationalization Association

Rationalization entailed the creation of a new *Selbstverwaltung* body, although this term was not employed. The Rationalization Association was delayed by difficult negotiations between the Federal and North-Rhine Westphalian governments over establishing the relative proportions of public finance which they were to provide, and the cold winter of 1962–3 somewhat eased the pressure to act. But such a body, furthering inter- and intra-company rationalization, was considered indispensable. German coal remained uncompetitive despite extensive improvements between 1957 and 1962 which involved eighteen pit closures, accompanied by an increase in labour productivity (OMS) from 1,651 kg. to 2,977 kg., with total employment falling from 345,000 to 256,000, and the proportion of coal mined entirely mechanically rising from 17 per cent to 60 per cent.[4]

With the pressure to rationalize thus continuing, the government announced the following 'Guidelines regarding preliminary granting of "premiums" for the closure of hard-coal mines of 13 February 1962'. Their purpose was both to increase efficiency and competitiveness and to speed up adaptation to declining market opportunities, for which public means would be made available towards necessary pit closures. The public purse was to provide the sum of 12.50 DM per ton of lost capacity, an amount doubled by the mining companies themselves, using their own resources. This promotion of negative rationalization was to apply to closures effected between 15 May 1962 and 30 June 1964 (provided notification of the closures and appropriate compensation claims had been lodged before the Rationalization Act came into effect in the autumn of

1963). As a direct result of these measures twelve mining companies with a total production of 7.9 million tons were closed down within fifteen months. In total, this 'preliminary action' (*Vorausaktion*) saw twenty-three cases of 'premium' awards, totalling 99 million DM.

The Rationalization Association was finally created by Act of Parliament on 1 September 1963. It was intended to facilitate the development of a collective view of how best to enhance the competitiveness of the whole industry,[5] and in this spirit the association ensured a unified response. The common purpose outweighed the individualism of company egocentricity. Such an organization did not represent a major innovation, it was merely a modern variation on a very old theme, industry-wide co-ordination on the one hand, and state–industry co-operation on the other. The sheer smoothness and effectiveness of this concertation was to evidence that blurring of the distinction between the state and industry which had characterized the pre-war period. Specifically, the purpose was to proceed beyond the negative rationalization of pit closures. The Act encouraged the amalgamation of coal-mines and the exploitation of adjacent coal lots, this was also facilitated by the general promotion of their sale, exchange, or leasing; it also envisaged acquiring shares in particular companies. All of this was intended to promote the creation of economically viable mining companies and the Federal government was prepared to continue the closure grants of 12.50 DM per ton (again, doubled by the industry itself), the public means being provided by income from the heating oil tax.

The Rationalization Association was empowered to grant loans and credit, on the basis of Federal guarantees, up to a total of 1.5 billion DM. The 25 DM closure premium itself was a basic grant, and additional amounts could be awarded in special cases, although no one company would be allowed to receive more than 100 million DM. The mineworkers strongly disapproved of the association; in the words of their leader Heinrich Gutermuth it was a 'pit-death association' ('Zechensterbeverein'). In particular, the miners deplored the secrecy in which the association conducted its affairs, for it was not required to publish the names of companies or mines applying for liquidation assistance, a situation which exacerbated mineworkers' sense of insecurity. The employers insisted on secrecy, however, because once miners learned that their pit was likely to be doomed they left promptly to seek alternative employment, hastening the date of closure. But, on the whole, there seems little doubt that the Rationalization Association was conceived and constructed with little regard to the workers' interests, reinforcing the demands of some for socialization on the basis of common ownership, as this was apparently the best way of preventing the association from assuming the role of funeral director (*Beerdigungsinstitut*).

The Rationalization Act had presented the association as a 'Federal public-law organization', a 'bundesunmittelbare Körperschaft des öffentlichen Rechts'. This was justified in the Act itself as an antidote to the possibly adverse economic, social, and political consequences of an over-hasty adaptation of coal-mining to the changing energy market, a legitimate public responsibility to be exercised by the public administration. So the association was a state organ fulfilling a public purpose, yet, according to one observer, the ultimate decisions (regarding closures) lay with the entrepreneur:

The selection of plant for closure remains ... the prerogative of the mining companies because the decision to close or continue working a pit is a matter for the owner ... Recognizing this, the legislator has refrained from limiting the powers of the employer in closing or continuing to run his works.[6]

This apparent contradiction between the precise responsibilities of the state and the private entrepreneur can only be reconciled in practice on the basis of a *Selbstverwaltung* compatible with a long German tradition.

Paradoxically, to a certain extent the very success of the Notgemeinschaft and the Rationalization Association was proving to be counter-productive. The former's success, in conjunction with the coal import tariff, not only shielded German coal from third-country competition, it shielded domestic coal's competitors oil and gas. The Rationalization Association's success in promoting negative and positive rationalization resulted in production being concentrated on the more efficient pits. This enabled high levels of output consistent with the overall output target of 140 million tons annually; but such technically efficient production was still high-cost production on an international comparison as coal was being produced far in excess of market requirements. This made it increasingly necessary to close down large and often relatively new and productive pits (as with the projected Graf Bismarck closure), shattering the morale of mineworkers and their families and threatening catastrophe for traditional mining communities which were almost entirely dependent on coal-mining to maintain their economic and social fabric. The realities of the energy market were moving strongly against German coal, the government's policy was in shreds, and there was a developing government crisis, with the national economy clearly moving into recession. All this foretold that a new and more ominous coal crisis was emerging, but was a largely private, industry-only solution capable of dealing with the scale of the problem, especially when the 140 million tons target had to be abandoned? This is where the corporatist nature of German political culture was to come to the assistance of the coal industry, adding a new dimension to coal *Selbstverwaltung*.

The Community Action for German Hard-Coal Mining Regions

In response to the deepening structural crisis, it was Fritz Berg, influential head of the Federation of German Industry,[7] who called a meeting on 27 April 1966 of top representatives from industry and from the governments involved in coal's travails. The meeting established a committee to prepare the way for the creation of the Community Action for German Hard-Coal Mining Regions Co. Ltd. (Aktionsgemeinschaft Deutsche Steinkohlenreviere GmbH). Its primary purpose was to manage pit closures in such a way that labour would only be shed compatible with the ability of the local labour market to absorb redundant coal-miners.[8]

It was initially conceived that a basic capital of 200 million DM would be provided by about one hundred companies, with Community Action granting loans up to a limit of 400 million DM, the sums involved being guaranteed by the state ('die öffentliche Hand'). Companies subscribing the initial capital would be immediately entitled to special depreciation allowances. Pit closures would be subsidized by state grants (premiums) of 15 DM per ton; the additional 8 DM per ton Equalization of Burdens levy would be waived. The vital matter of acquiring industrial land for new industries was to be tackled by offering mining companies 30 DM per square metre, certainly not an unprofitable transaction seeing that many companies' land was registered at a book value of only 50 pfennigs. However, should reticence in this matter seem a handicap, Community Action was able to refuse to pay the 15 DM per ton closure premium if the whole site were not to be offered as part of the closure deal. (This unified control of land flow and industrial sites also permitted the warding off of unwelcome foreign intrusion.) Mobilizing site availability, while limiting speculation, was a major priority. Although not initially planned, it was later conceded, in 1968, that local authorities would be entitled to purchase land becoming available. Community Action was freed from subsidence liabilities, through state guarantees, and from related tax liabilities.

However, its administrative council was not formed for six months, until 28 October 1966. When Community Action itself was formally established on 23 November 1966, only 65 million DM basic capital had been raised, and that from 500 companies! It seems that the continuing government crisis, the deepening recession, and talk of a fundamental reorganization of the whole industry had created considerable insecurity in business circles, making the implications of involvement very uncertain. It was nearly a further three months before the Bundestag enacted the necessary legislation regarding the relevant tax privileges (15 February 1967). This was endorsed by the Bundesrat on 3 March 1967. Community Action, as a non-profit-making body, was exempted from the corporation, trade, wealth, and company taxes.

Finally, a year after Berg's original initiative, an Agreement was signed between the Federal government and Community Action, including Guidelines on Granting Closure Premiums on 22 March 1967. By the end of 1968, Community Action had organized seventeen pit closures, entailing the abandonment of 16.8 million tons of annual coal capacity.

It is quite paradoxical that the first two of the three major organizations created to combat specific aspects of German coal's crisis, the Notgemeinschaft and the Rationalization Association, were founded on state initiative, yet staffed entirely by private organizations, admittedly fulfilling a public purpose, whereas the third organization, Community Action, sponsored primarily by private initiative, was one in which public representatives were prominent. Community Action's administrative council included the Ministers of Economics from North-Rhine Westphalia and the Saar, Dr Fritz Kassman and Dr Reinhard Koch, while the Federal government was represented by the Coal Commissioner, Dr Gerhard Woratz, formerly head of Department III in the Federal Economics Ministry. Other members indicating a public interest were Dr Alfred Einnatz of the RWE; Dr Heinz Kemper of VEBA; and Dr Heinz Sippel of the West German State Bank Giro Centre, Düsseldorf. The public presence was even more evident in the advisory council.

Employers were also substantially represented on the administrative council: one of the two deputy chairmen was Fritz Berg himself; also members were Otto Wolf von Amerongen, and Dr Helmut Burckhardt of the Coal Employers. All these leading private and public representatives were united in a concerted action 'to ease the orderly closure of coal-mines and to improve the economic structure of the German hard-coal mining districts'.[9] Five hundred and fifteen private companies were attempting to co-ordinate measures with each other and with the state in a vast, co-operative enterprise, revealing both a great sense of social responsibility and a desire to obviate direct and detailed government intervention and regulation. This venture can be described as corporatist in spirit and in practice, demonstrating also how flexible a phenomenon *Selbstverwaltung* is.

Each of the self-help organizations described above represented a more emphatic response to the deepening coal problem as perceptions widened as to the real nature of the predicament. In the late 1950s it was assumed that coal's difficulties were temporary and the Notgemeinschaft was created to fulfil a limited holding operation, to keep third-country coal off the German market, to give the domestic coal industry time to respond to this sudden foreign challenge. But by the early 1960s, the structural nature of coal's travails was becoming more obvious, not least through the penetration of fuel markets by cheap foreign oil, so the Rationalization Association was formed, with the specific purpose of forcing rationalization in colliery organization and equipment, as this was apparently the only effective means of responding to coal's multifarious challengers. By the mid-1960s it was appreciated how the coal industry's economic crisis was embedded in an incipient social crisis, as the very existence of whole *montan*-communities was threatened with imminent disaster. The creation of Community Action, with its objective of an immediate and thoroughgoing industrial diversification, was an expression of the seriousness and complexity of such a threatening coal catastrophe.

This impending socio-economic crisis in Germany's industrial heartland necessarily presented a political crisis going beyond technical questions of policy design. To use an Americanism, the whole West German political economy and its newly won democracy were on the line. Failure to act promptly and effectively, within the context of a general recession, would nurture the enticements of incipient political extremism, with the danger of the experience of the late 1920s repeating itself. This was the nature of the national crisis which emerged, favouring the rallying of all democratic political forces, to ensure the survival and stability of the existing system. At the national level, it was given practical expression in the creation of the Grand Coalition.

However much opinions may be divided on the relative merits of the Coalition it was still a harsh fact that coal's crisis was to become relieved and postponed, not resolved. As has been demonstrated, Community Action could not really commence effective action until 1968. It was state action, sometimes euphemistically characterized as flanking measures, which created the indispensable context for the functioning of the latest and last self-help, self-regulatory organization, the Ruhrkohle itself. But how precisely was it self-regulatory? It has been established that the autonomy of German industrial self-regulation is a very relative notion. Formal mechanisms tell, at best, only half the story.

The Ruhrkohle AG

This company, which owed its very existence to appropriate legislation and public financial support, was neither publicly owned nor was there an overall majority or minority state shareholding. The mainly private shareholders were the *Altgesellschaften* (old companies). The company's leading personnel came from the industry and the central coal organizations provided five of the six members of the Ruhrkohle's board of management. The umbrella organization of German coal employers (the Gesamtverband) provided the chairman, Dr Hans-Helmut Kuhnke, and Friedrich Carl Erasmus, both from the Gesamtverband's own board of management; the Ruhr Coal Employers (UVR) sent Hubert Grunewald, again from its own board of management. Other members were Karl Heinz Hawner, chairman of the industry's technical association the Steinkohlenbergbauverein, and Ernst Schmid, head of the Präsident coal sales agency. The sixth member, as labour director, was Heinz Kegel, formerly of the executive committee of IG Bergbau (until 1965) and SPD Member of the North-Rhine Westphalian *Landtag* (until 1966). The German coal industry itself was clearly capable of managing its own affairs in a most direct way.

The coal and steel employers and the mining union also exerted, as was to be expected, a dominant influence on the twenty-one-member supervisory board. Its chairman was managing director Dr Heinz P. Kemper, chairman of the boards of management of both the Gesamtverband and of the UVR; first deputy-chairman was Adolf Schmidt, MdB, head of the mineworkers' union; and one of the further two deputy-chairmen was Dr Hans Günther Sohl. Schmidt was closely supported by Heinz-Werner Meyer of IG Bergbau's executive committee and later MdL (from 1975), and by Alfons Lappas, from the executive committee of the DGB; also important was the presence of Walter Hesselbach, as head of the Bank für Gemeinwirtschaft, a trade union representative.[10] There was, in effect, a most remarkable criss-crossing of interests: regional, industrial, and political of which the supervisory board of the Ruhrkohle AG was a microcosm.

But where did power lie? Presumably with the most dominant groups on the company's supervisory board which, after all, appointed the board of management made up of top executives from within the industry.

There was no perceived need to bring in outsiders, such as a former merchant banker or, least of all, executives from abroad. So in one technical sense the industry was perfectly self-regulatory, providing its own managerial expertise at the highest level. Nevertheless, as the history of the previous fifty years had demonstrated, there could be no pure coal interest equal to, let alone superordinate to, certain other interests. The traditional steel influence remained important, with a more recent influence exerted by the semi-state (mainly energy) conglomerate VEBA, at that time the largest industrial undertaking in the Federal Republic. If there was a pure coal interest then it was more likely to be expressed by labour and its vehicle of formulation, IG Bergbau und Energie. By the early 1980s, this by then relatively small union possessed a total of 156 seats on various supervisory boards (79 directly and 77 indirectly), 74 being on the boards of ten established *montan*-companies. The 1970s and after were also to reveal a fascinating fluctuation in influence and dominance between VEBA, with its oil and electricity interests, and the steel barons.

But perhaps of more interest, how precisely did the manifest public interest and stake in the coal industry express itself? How did the state ensure that the sums it invested were appropriately applied? These sums were substantial. For example, the Federal and two *Land* governments (Saar and NRW) were providing by the end of the decade, in 1980 alone, 4.46 billion DM (excluding the 2 billion DM coal penny). The state-owned Saar Mining Co., the semi-state VEBA, and the local authority-dominated RWE represented a very amorphous and ill-co-ordinated public sector. It is tempting to want to investigate the reality of state control through such formal mechanisms as existing companies, but no direct chain of command ran directly from, say, the Federal Economics Ministry through to the operations in the field of individual companies; no elaborate planning bodies existed to balance and direct relevant strategies in the interconnected industries of coal, iron and steel, electricity generation, and chemicals.

This is not to deny the important role VEBA occupied within government energy policy, but direct state intervention would have breached a long-standing official consensus that civil servants cannot run industry. Direct intervention would have offended prevailing political ideologies and repelled executives of the calibre and standing of Bennigsen-Förder, the contemporaneous chairman of VEBA's board of management who would not have wished to tolerate detailed state interference. Informal channels of influence and guidance are the primary means of state or public involvement, and nowhere was this more evident than in the organization and financing of the crucial and expensive total social plan, or, more revealingly, of the adaptation money (*Anpassungsgeld*).[11]

But what were the fundamental characteristics underpinning the whole adaptation money scheme, now that, since the formation of the Ruhrkohle AG, a non-labour perspective on self-regulation had become virtually irrelevant? They were threefold, and in the view of the findings already established, quite predictable. First, the influence of the mining union, IG Bergbau und Energie. The very scheme in question resulted from the union's initiative. This comprehensive and generous scheme was unique, no other group of German workers had attained a comparable level of social security. The second characteristic was the dominant and decisive role of the state. Quite apart from the elaborate and detailed corpus of legal provisions and administrative regulations which only the state could provide, the central issue of finance was in this case, by definition, a publicly funded responsibility. Above all, the ultimate authority of the Federal Coal Commissioner in matters of adaptation money (as in all adaptation measures) was final. (It is difficult to conceive how British-style nationalization could grant government more influence in directing coal affairs.) Thirdly, this state influence was both assisted and modified by the remarkably high level of co-operation, co-ordination, indeed integration, of the relevant interested parties: state, employers, and labour consciously pursued an active policy of concertation. Perhaps the most outstanding feature of the whole phenomenon of the adaptation money scheme was the centrality of the state role and the largely informal manner in which this was fulfilled. A fascinating paradox obtains in the contrast with Germany's apparently legalistic political tradition, clearly suggesting, if not demonstrating, the subordinate significance of legislative provision in the solution of politico-economic crises.

So clarity is required on the precise meaning of the terms 'formality' and 'informality' in West German politics. In a trivial sense crisis management entails a formal element in that some kind of legislative enactment is required, but the precise form thereof and its interpretation is likely to be very open and extremely pragmatic. Moreover, there seems to be little evidence of formality in the sense of institutionalizing state–industry interaction in particular joint organizations, meetings appeared to occur largely on an *ad hoc* basis and, of course, invariably in private. Institutions created by the Coal Act, other than the Office of the Federal Coal Commissioner, generally kept a low profile. Bilateral and multilateral negotiations generally took place behind closed doors and, in the case of the coal industry, produced workable solutions: there were few strikes, lock-outs, or even investment strikes—demonstrations were normally peaceful and orderly.

In practice, conflict was minimized and interest equalization maximized. One means whereby this was achieved was on the supervisory board of a company like the Ruhrkohle AG where public, labour, and private representatives from various associated industries were in constant contact and information flow and personal interaction were maintained in such a way as to build confidence and facilitate agreement in cases where interests clashed. But organizations seldom seem to create values; as the British National Economic Development Council amply confirmed, they can at best provide a framework within which new values may emerge or provide a channel for existing values and attitudes to manifest themselves explicitly. This is the great strength of the West German institution of co-determination. It cannot cause agreement between intransigent antagonists, but it can provide a reliable framework for reconciliation where the will to compromise exists.

The ubiquitous informality outlined above can lend itself to the criticism of excessive confidentiality whereby deals are arrived at by powerful interest groups, to the disadvantage of the public at large, especially in the case of coal-mining where huge sums of public money are involved, both for employers, e.g. closure grants, and for miners, e.g. adaptation money. Is this a classic illustration of employers and employees combining in a conspiracy at the expense of the taxpayer? Not necessarily. One essential component of the dominant role of the state in industry is that of independent arbitrator, a role which possesses a long tradition in Germany and wide acceptance within West German political culture, whereby the state is perceived as the active guardian of the public interest. The system as it has evolved since the Second World War has certainly spared the Federal Republic major conflicts and dislocations of the kind which have often wracked Britain, France, and Italy. But the German system of political economy is not necessarily readily transferable to other societies; it is very much the product of Germany's own historico-geographical circumstances.

The Labour Dimension

In the period following the Second World War labour participation in industrial self-government became intimately associated with the establishment and functioning of the institution of co-determination, primarily at company level in the *montan*-sector, and primarily at works council level in other industries. *Montan*-co-determination, or parity co-determination (or qualified co-determination), was characterized by parity representation of capital and labour on company supervisory boards with a neutral eleventh man. The labour director on the board of management was responsible for social and personnel affairs, and since he could not, in practice, be appointed against the wishes of labour representatives, he became a labour nominee. *Montan*-co-determination was originally established in March 1947 in the British occupation zone in the first decartelized steel companies. By 1950 this form of co-determination had by no means been extended to the whole of the steel industry and certainly not to coal, except in those companies entirely owned by steel concerns. Indeed, it was only after a major domestic political battle, when the IG Metall and IG Bergbau held strike ballots which yielded pro-strike results of over 90 per cent, that Federal legislation was enacted, in 1951, firmly and legally establishing *montan*-co-determination for both industries.

However, it would certainly be inappropriate to characterize IG Bergbau as steadfastly pursuing the introduction of parity co-determination from a very early date after the close of the Second World War. It would seem that in practice three priorities emerged for the union after the war,[12] in the following order of importance: first, a sensible and efficient scheme of industry-wide reorganization and planning for coal;[13] secondly, as a means to that end, socialization through common ownership; and, thirdly, co-determination of labour through the union. The actual implementation of parity co-determination came to represent something of an important fall-back position, one which could be rationalized in terms of economic democracy, but was as much a matter both of individual trade union power and pragmatic industrial management.

The Mining Union and Coal Industry Reorganization

The union appeared to follow a consistent line regarding the socialization of the coal industry for a period of more than fifteen years. An initial exposition of this line was presented at the union's first general meeting on 8 December 1946 by Ludwig Lehmann,[14] formerly of the executive committee of the Alter Verband (1930–3). A mere month later, Dr Georg Berger, head of the union's economics department, presented a much more systematic and detailed proposal still;[15] socialization here is clearly equated with common ownership. A resolution was passed five months later at the Second Interzonal Conference of the Mining Union(s),[16] very similar in content to Berger's proposal, demanding the transfer of mining companies into common ownership with the trade unions enjoying full equality. However, from the autumn of 1947 onwards Allied measures to increase coal production by offering material inducements to miners, in terms, for instance, of extra rations, drove something of a wedge between them and their union, with the union's demand for socialization appearing less relevant as coal production actually rose with each new incentive.[17] With the announcement of the Marshall Plan and the associated ascendancy of American influence, particularly through 1948, hopes of socialization became increasingly unrealistic.

Yet, repeatedly in 1948, the union continued to maintain its demand for socialization of the coal industry as, for example, when announcing its support for the Marshall Plan on 15 April 1948. In May 1948 Heinrich Jochem, also a member of the NRW *Landtag*, renewed the call for socialization and, only one and a half months later, Gutermuth, in an appeal to miners to increase production, reiterated the objective of socialization but emphasized that much public opinion was against socialization which now appeared to have become a longer-term goal. None the less, on 23 October 1948, at a Ruhr conference of works councillors, both the objective of socialization and of 'a decisive right of co-determination' were reaffirmed.

Yet despite the apparent consistency of the union's formal demands, there were legitimate doubts as to its ability to pursue them forcefully because of diversions within its own ranks. This was apparent in the *Landtag* debates of 1948 when the union, this delicately balanced *Einheitsgewerkschaft*, could not insist on a common line in the debates on coal socialization.[18] It was evident again in the union's Memorandum of 5 April 1949, regarding Law No. 75, which mentioned virtually nothing about reorganization and co-determination. The list of demands, including a board of coal industry trustees with parity representation with which the union finally responded in its 2nd Memorandum, of 29 April 1949, had very clearly been settled in advance, after discussions with the DGB and IG Metall.

Indeed, it appeared to have been the latter two organizations which stiffened IG Bergbau's resolve in the two vital areas of co-determination and common ownership. These demands were added retrospectively to the mining union's full acceptance of the DKBL's reorganization plan for the coal industry which appeared in September 1950, calling for twenty-three new coal companies to be created. Even more importantly, after news from the Economics Ministry in the summer of 1950 that parity co-determination was to be withdrawn in the steel industry, the fight to maintain it was borne primarily by IG Metall and the DGB, particularly the latter, with Hans Böckler succeeding in getting parity co-determination extended to the coal industry, against Heinrich Kost's bitter opposition. IG Bergbau was content with supporting the initiatives and efforts of the other two trade union organizations.

The appearance finally in 1951 of the *Montan* Co-determination Act on the statute book by no means signalled the end of disputes either on issues of principle or of implementation,[19] with Mannesmann leading employer opposition.[20] The core of the continuing controversy was the Allied failure to promote effective decartelization and the emergence of mightier-than-ever *montan*-combines, representing a threat to the future of co-determination. The passing of the second Act in 1956, however, did not obviate, in the mining union's view, the need for a complete reorganization of the industry, as confirmed at an extraordinary General Meeting of the union in May 1956.

When the first coal crisis broke two years later, the union felt not only that the correctness of its basic position had been confirmed, but that it was also compelled to add to its standing demands for common ownership of the coal industry and for a Coal Council (a form of extra-enterprise co-determination) the demands both for an Energy Council and for an appropriate energy policy. Only common ownership, apparently, could guarantee the creation of coal companies of an optimal size to combat the prevailing crisis.[21] Seven months later, in January 1959, the union also called for the creation of the post of Energy Minister. But with the apparent passing of the first coal crisis within a further eighteen months all these demands appeared to lose their relevance.[22]

Some First Practical Consequences of Montan *Co-determination*

The *Montan* Co-determination Act of 1951 had formally integrated mining labour into the senior management structures of the coal industry, but had parity co-determination contributed towards moderating the hundred-year-old coal employers' tradition of implacable hostility towards organized labour? Blume *et al.*'s authoritative review of the practice of *montan* co-determination after the first ten years offers a valuable insight into this question.[23]

Comparisons were made between experience in the steel and coal industries, with findings generally to the detriment of coal. For example, instances where the labour director (*Arbeitsdirektor*) also determined company pay policy for middle management were almost only to be found in the iron and steel industry. Cases of joint decisions between labour director and the director of the area desiring recruitment (commercial or technical) again were usually to be found only in iron and steel. Generally, labour directors were much more active in employers' associations in the steel industry than in the coal industry. Labour directors in iron and steel were much more likely both to represent their company in the relevant insurance associations and to act for it in the subsidiary companies.[24]

The *Zwischenbilanz* study finally highlights, more objectively, two important constraints on labour directors in coal-mining which were much less in evidence in the steel industry, those of resources, especially staffing, and of company structure. Labour directors in iron and steel possessed more staff to meet the demands made upon them; there were major coal companies whose labour directors did not even possess a secretary. Generally, the inferior contact of labour directors with their board of management colleagues was explained by the authoritarian form of company organization in mining.[25] Given the key role allotted to the labour director within the whole institution of company (*montan*) co-determination,[26] the prospects for meaningful teamwork between management and labour in a crisis did not appear at all favourable.

Labour Co-determination and the First Coal Crisis

So, what was the precise impact of the first coal crisis on management–labour relations? How relevant were the existing co-determination arrangements in preventing a polarization of attitudes and action and in enhancing the creation of a common front? One study, conducted in the mid-1960s,[27] indicated that co-operative agreements normally prevailed over protracted, disputatious conflict and that the labour director was permitted, indeed forced, to play an active role in closure and rationalization negotiations with the relevant work-forces, a situation where there was more conflict potential with these workers than with his managerial colleagues. Generally, the labour director was to play a very constructive part in conflict management.[28]

The main enduring result of the heightened co-determination activity during the crisis, in terms of practical innovation, were the new social plans. This development was very much furthered by the union, although in each specific case it was generally the labour director who took the initiative, not the employer or the works council. He was in a strong position to do so because of the board of management's desire for his support over closures. The importance of the negotiations can be gauged by the fact that the works council was often led in the negotiations by a member of the executive committee of the union. In the early years of the crisis it was common for the union to present a ready-made plan for acceptance; with the passage of time the works council began to assert increasing independence in the negotiations.[29]

The social plan was consistent with a posture of generosity towards the work-force. As regards transferring miners:

Although the transfer seldom led to difficulties of adaptation, the employee enjoyed in numerous cases a payment agreed in the social plan which then led to a higher income than before the transfer.[30]

Even more significantly, regarding miners not transferred:

In such a way, the income of an unemployed former miner could be higher than that of a re-employed former miner if the latter received a benefit or benefit-like payment as well as a works pension or a transitional payment which he lost as soon as he took up more than minor employment/according to the rules of the pension scheme.[31]

So it was not just as a result of arrangements made during the second coal crisis that many miners ended up better off out of employment than in; the precedent had already been established.

Although it is to be noted that constant change in the ECSC regulations often had disadvantageous effects on social plans for miners, it would appear that, in general, works, Federal, and supranational schemes complemented each other well. For example, most (works) social plans envisaged advances to miners until the other schemes actually came into effect. Numerous social plans were amended retrospectively, to the advantage of miners. The achievements of works social plans represented a success for plant-level co-determination on which they were entirely dependent.

But how did the co-determination legislation of 1951 materially affect crisis management? In non-labour circles the position of labour director had been much maligned as an anomalous and incongruous invention; yet during the first coal crisis it definitely proved its value to both sides of the industry, substantially improving social provision for miners and minimizing debilitating conflict, to management's advantage. Practice appeared to refute theory here. The supervisory board which, according to official labour opinion, was the linchpin of company co-determination evinced the most evident contradictions when subject to acute pressure and, as in the case of closures, seemed to represent the least satisfactory of the organs of co-determination. The one outstanding success of co-determination had been at plant level, with the works councils, but these had not even been included in the relevant *montan* co-determination legislation.

A spontaneous importance accrued to the least likely organ, the works council. Yet in theory this institution represented difficulties for the union both because of legal constraints on the councils (through the 1952 Act, requiring non-partisan approaches) and in practical terms as they were not part of the union machine. However, the union's exceptionally high degree of organization, compared both with the pre-war years and with other contemporary trade unions, gave the mining union a degree of influence on the labour side which facilitated a co-ordination of effort which was not likely to be matched by many work-forces in other industries. Certainly, the lessons regarding the necessity to co-ordinate activity between the three primary organs of co-determination were not lost on the union. It was the only organization able and willing to make the links and would, with time, prove to be equal to the task, whether from a national or trade union perspective. It would seem, then, that the main value of the superordinate co-determination organs had been as a strategic back-up for the works council.[32]

The development of the social plans and also the ensuing greater understanding displayed in collective bargaining represented significant movement by the employers regarding the institution of co-determination. It could be interpreted as their contribution towards a trend of convergence between capital and labour over management of the industry as a whole, for by the mid-1960s IG Bergbau had significantly shifted its position over the question of ownership, hitherto considered crucial to meaningful coal industry reorganization. As recently as 22 July 1963, the union had reiterated the importance of common ownership in its restatement of energy principles. However, 1964 revealed that the union was no longer considering common ownership to be a prerequisite of a fundamental reorganization. In 1965, at a miners' conference in Oberhausen, Kegel confirmed that the union's latest proposals for the industry were 'beyond ideology'.[33] A clear line had been drawn between property relations and reorganization, with the latter enjoying priority in pursuit of the pragmatic objective of a new order for mining and the energy sector. On 19 December 1966, the mining union presented its own model bill for the complete reorganization of the West German coal industry, containing a plan which envisaged a private solution; public ownership was definitely now a thing of the past.

The Establishment of the Ruhrkohle AG and the Entrenchment of Co-determination

Once the second coal crisis broke, negotiations governing the reorganization of the industry in general and co-determination in particular were to prove both long and hard, not only because of their inevitable complexity but also because the basic Rheinstahl plan had not envisaged continuance of parity co-determination in the company organs. At one point, on 6 February 1968, IG Bergbau informed the Economics Minister that it considered further negotiations of the prevailing nature pointless and accordingly planned to withdraw completely.[34] Schiller initiated a new series of negotiations and to ensure IG Bergbau participation he informed the latter that henceforth negotiations would no longer proceed on the basis of the Rheinstahl circle proposals. One of the first and main results of this fresh round of negotiations was agreement on, and declaration of, the 'Total Social Plan regarding public and company payments and preliminary measures for employees affected by closures' on 15 May 1968.[35] The creation of the one unitary plan both foreshadowed and was a prerequisite for the establishment of one major coal company to control the Ruhr, which covered 75 per cent of all West German production, although, of course, the plan also applied in the other mining areas.

At the decisive meeting on 14 June 1968, the union managed to obtain in the Preamble to the Basic Contract, which had been agreed between the Federal government, the parent companies, and the Ruhrkohle AG, the following co-determination regulation:

1. The statutory co-determination of the *montan*-industry applies to the total company, and also at group level.
2. In the works units a plant director for personnel and social questions will be installed, subordinate to the works technical director, but in other respects superior to other offices in his own sphere.
    This plant director will be appointed by the board of management of the appropriate group, upon nomination by the labour director.[36]

This regulation formed the basis of the final, formal agreement between the executive of IG Bergbau and the board of management of the Ruhrkohle AG, issued on 18 July 1969. The agreement did not match in full the hopes of the union, there had to be compromises. Nevertheless, it was finally stipulated that the works director, although normally subordinate

Bahl, *Staatliche Politik am Beispiel der Kohle* (Frankfurt: Campus Verlag, 1977), 113); given its contemporary importance as a democratic stabilizer and its enduring crucial and indispensable role as ersatz-ideology for the German labour movement the foregoing observation seems incredibly inappropriate.

3. '. . . the danger of a further submersion of the German market by imported coal was fended off in a little under two years by an appropriate co-operation between the Federal government and coal-mining, between government economic policy measures and employers' self-help'; quoted in U. Specht, 'Die Energiepolitik der Bundesrepublik von 1948–1967', Ph.D. thesis, Albert-Ludwigs-Universität, Freiburg, 1969, 111–12.

4. The report of the special independent Energy Inquiry established by parliament (the *Energiegutachten*) had predicted rising labour costs but constant company returns, the whole problem being exacerbated by a shortage of mining recruits (ibid. 113).

5. See G. Krink, 'Die Energiepolitischen Maßnahmen der Bundesregierung', in H. Schmidt (ed.), *Energiewirtschaft und Energiepolitik in der Gegenwart* (Berlin: Duncker & Humblot, 1966), 176.

6. C. D. Schmidt, 'Die Krise im Steinkohlenbergbau und ihre soziale Problematik unter besonderer Berücksichtigung des Ruhrgebietes', Ph.D. thesis, Westphalian Wilhelms-Universität, Münster, 1967, 325.

7. In the autumn of 1960 Berg claimed that he only needed to call on the Federal Chancellor and the government would drop any proposal in question (by no means an isolated incident). See T. Eschenburg, *Zur politischen Praxis in der Bundesrepublik* (Munich: R. Piper & Co. Verlag, 1964), 240.

8. But equally, it had an important role in job creation, encouraging existing non-coal employers to expand and, above all, to attract new industries to the coal-mining areas.

9. *Jahrbuch für Bergbau, Energie, Mineralöl und Chemie 1970* (Essen: Glückauf, 1970), 653.

10. Adolf Schmidt himself later became deputy-chairman of this trade union bank's supervisory board. He was later succeeded by Meyer as head of the union; Meyer later became head of DGB, and Hans Berger became head of IGBE. Heinz Horn became chief executive of the Ruhrkohle AG.

11. For a detailed account of the scheme see H. Zydek, 'Anpassungsgeld: Ein Beitrag zur Entstehungsgeschichte, zum Inhalt und zur Auslegung der Anpassungsregelung', *Der Kompass*, 1 (1972).

12. According to M. Martiny, 'Die Durchsetzung der Mitbestimmung im deutschen Bergbau', in U. Borsdorf and H. Mommsen (eds.), *Glück auf, Kameraden! Die Bergarbeiter und ihre Organisationen in Deutschland* (Cologne: Bund-Verlag, 1979), 401.

13. See e.g. Document 22 in M. Martiny and H. J. Schneider (eds.), *Deutsche Energiepolitik seit 1945* (Cologne: Bund-Verlag, 1981), 88.

14. His main recommendation was a partial nationalization of the coal industry, on the basis of a co-operative (*Genossenschaft*); this was not full nationalization, with its danger of bureaucratization, but the state position had to reflect a first among equals status because of its role as a guardian of the national interest (*Gesamtinteresse*), and as the active integrator of conflicting groups and interests. For further details, see Document 1, ibid. 21–4.

15. Advocated was a Central Office for Mining (modelled on the Imperial Coal Council), an office in which the mining union would be directly involved, as also in any 'democratic co-determination co-ordinated by this administration' (Document 2, ibid. 25–33). The basic company unit would be a *Sozialgewerkschaft*; these units would be co-ordinated by a board of trustees (*Treuhänderkollegium*) on behalf of the German people ('das deutsche Volk in seiner Gesamtheit'), as the bearer of the legal title to the mining property (ibid. 29–30).

16. See Document 3, ibid. 34–6.

17. See U. Borsdorf, 'Speck oder Sozialisierung', in Borsdorf and Mommsen (eds.), *Glück auf, Kameraden!*, 366.

18. See W. Rudzio, 'Das Ringen um die Sozialisierung der Kohlewirtschaft nach dem Zweiten Weltkrieg', in Borsdorf and Mommsen (eds.), *Glück auf, Kameraden!*, 383.

19. For details, see H. Deist, *Die Neuordnung in der Montanwirtschaft in den Holding-Gesellschaften* (Bochum: IG Bergbau, 1955).

20. In response to provocation by the leading industrialist Reusch, in particular his flat rejection of the proposed Co-determination Extension Bill, the two main *montan*-unions sanctioned a 24-hour protest strike on Saturday, 22 Jan. 1955. See W. Beer, *Mitbestimmung* (Bochum: IG Bergbau, 1975); more than 800,000 workers went on strike, nearly 600,000 of them miners.

21. The Federal government would hold 74% of the shares in the relevant holding company, the *Länder* holding the remainder; parity representation of labour on the supervisory board would be provided entirely by the union itself, Martiny and Schneider (eds.), *Deutsche Energiepolitik*, Document 29, 134.

22. Nevertheless, as the new decade unfolded, IG Bergbau appeared to be the only organization possessing a comprehensive energy plan of its own (ibid. 158), further evidence of its invariably constructive approach, giving the union a certain moral pre-eminence and enhancing its right to be heard when the next, more serious crisis broke.

23. O. Blume, H. Duvernell, and E. Potthoff, *Zwischenbilanz der Mitbestimmung* (Tübingen: J. C. B. Mohr (Paul Siebeck), 1962).

24. The quality of social relationships at board of management level in mining had changed little—mostly a bleak picture obtained in the industry: 'You can't possibly have anything to do with such people' (ibid. 115); private contact was impossible because of the allegedly unbearable arrogance of the *Bergassessoren* (ibid. 142 and 164).

25. Ibid. 272.
26. See not only ibid. 98, but also Shuchmann who had predicted that 'the ultimate success or failure of co-determination within enterprise depends largely on the ability of the labour director to foster among the workers the feeling that they are partners in a common enterprise' (A. Shuchmann, *Co-determination: Labour's Middle Way in Germany* (Washington, DC: Public Affairs Press, 1957), 154).
27. C. D. Schmidt, 'Die Krise im Steinkohlenbergbau und ihre soziale Problematik unter besonderer Berücksichtigung des Ruhrgebietes', Ph.D. thesis, Westfälische Wilhelms-Universität, Münster, 1967.
28. But although such an active role was structurally determined, the success of its conduct in terms of formal/informal co-determination depended to a great extent on personality factors (ibid. 383).
29. Some councils resented union tutelage and some labour directors disliked what they sometimes considered to be unrealistic demands (ibid. 457).
30. Ibid. 467.
31. Ibid. 471.
32. The practice of co-determination appeared also to have brought benefits in the sphere of collective bargaining. According to Schmidt, the wage negotiations of 1963 marked a watershed in employer–labour relations in the coal industry: their success was based, at least in no small part, on the precedents regarding the management of closures and the understanding and confidence created there. For confirmation of this, see ibid. 191.
33. See Martiny and Schneider (eds.), *Deutsche Energiepolitik*, Document 50, 258–9.
34. See W. Müller, 'Die Neuordnung des Ruhrkohlenbergbaus', *Das Mitbestimmungsgespräch*, 2 (1969), 24.
35. It consisted essentially of four parts: (*a*) the granting of a redundancy payment; (*b*) adaptation grants, half borne by the Federal government and half by the Commission of the European Community; (*c*) payments and other measures to be borne by the mining companies themselves; and (*d*) a final section elaborating in more detail the content of the preceding three sections. Reprinted in Martiny and Scheider (eds.), *Deutsche Energiepolitik*, 302–6.
36. Quoted in Beer, *Mitbestimmung*, 34.
37. In practical terms, the work of the works personnel director corresponds largely to that of the board of management's labour director; the former director cannot be appointed against the wishes of its union (ibid. 39).
38. 'The composition (of the supervisory board) guarantees that from the beginning of the Ruhrkohle AG onwards no decision will be taken against the will of employee representatives. . . ; it is not to be expected that decisions will be taken against the objections by trade union representatives' (Müller, 'Die Neuordnung des Ruhrkohlenbergbaus', 25).
39. Müller also implicitly acknowledges that coalition government had been crucial (ibid. 24–5), but that it was SPD participation which had proved so vital.

40. Just how important the union officially viewed its creation was revealed by Walter Arendt in a speech to the mineworkers conference on 29 Apr. 1969 when he compared it with the introduction of the direction principle (Martiny and Schneider (eds.), *Deutsche Energiepolitik*, Document 61, 324); this was a peculiarly apt analogy, for the direction principle had introduced state control of coal-mining, with state ownership playing an insignificant role.
41. Also a member of the SPD and later to become a Euro-MP.
42. W. Arendt, 'Mitbestimmung ist für die SPD paritätische Mitbestimmung', *Mitbestimmungsgespräch*, 11 (1970), 206. This was also confirmed in Vetter's speech (ibid. 207).
43. Martiny and Schneider (eds.), *Deutsche Energiepolitik*, 331.
44. Ibid. 327.
45. The social consequences of the projected adaptation programme, for example, were the subject of conversations between Adolf Schmidt and the Federal Chancellor on 30 Apr. 1971. For further details, see ibid. 327 and 341.
46. Ibid. 332.

9

Coal, Corporatism, and Labour

The active participation of organized labour in managing the affairs of the West German hard-coal industry, of the whole Ruhr economy itself, and, indirectly, of the Federal Republic does not represent an isolated phenomenon, an unprecedented performance by a unique organization. It can be argued that the union's activity merely demonstrates, in an advanced and compact manner, the propensities generally evident in the wider labour movement. Special factors influenced the precise form and nature of this union's contribution, but the basic thrust of its response and its ability to lead positively is characteristic of a constructive, conciliatory labour stance which was very much in evidence before the Second World War and, indeed, before the First World War. The inter-war and wartime experience was not to transform organized labour's approach to economic, political, and social organization; it merely reinforced the prevailing philosophy and attitudes and offered the opportunity to recognize and combat existing deficiencies and weaknesses, the primary feature of which had been internal division. The post-war environment facilitated consolidation and stability both internally and in the labour movement's relations with those other two pillars of modern industrial society: capital and the state.

The present chapter demonstrates how organized labour in post-war West Germany both embraced and was embraced by a corporatist system whose growth and stability has become so dependent upon the substantial breadth and depth of labour support that the latter has become a condition of system survival:

According to the corporatist model, the process of interest-reconciliation resembles not a free market but a kind of political cartel, which implies industrial self-government on the one hand, and the adoption of major interest-groups into the governmental process on the other. Corporatism also means the institutional growth, recognition and self-regulation of these major interests and, in our days especially, the triangular pattern of 'wheeling and dealing' between business associations, trade unions and the state.[1]

The Case of Coal and Electricity

The German coal industry represents one model of corporatism in a major industry. Clear precedents were established before the Second World War, with certain features and characteristics which were further developed and refined to become as much in evidence during the 1970s and even the 1980s as at any time previously. Nowhere is this phenomenon more apparent than in the coal industry's relations with electricity producers.

The pattern was demonstrated by the original coal–electricity agreement of early March 1977 which had amounted to a declaration of intent, followed by a formal agreement in May. The coal industry evinced an imbalance in its relations with the electricity supply industry. Just like the Ruhrkohle in the original Foundry Contract and Power Contract it emerged as the total debtor (*Gesamtschuldner*), reflecting its weakness in being so dependent on this particular market. On the other hand, it was backed by the authority of the state.[2] Indeed, the agreement between the utilities and the mining companies was only valid when 'the legal provisions necessary for its execution . . . are made'.[3]

The agreements had, in fact, only materialized because the state exerted pressure on the electricity companies and their associations, threatening direct intervention. Similarly, agreements would not have succeeded without a propensity for co-operation. Indeed, the whole episode was an eloquent testimony to the subordination of individualistic, competitive interests to the interests of a wider whole: self-interest was conceived in wide terms and best served by collective arrangements:[4] 'Agreement between utility and mining company is to be achieved by a system of co-operation and consultation.'[5]

The state had been called upon to reconcile two conflicting objectives: (i) 'assured supplies of low-priced energy are an essential element in maintaining the competitiveness of German industry';[6] and (ii) 'Priority use of domestic coal in supplying the Federal Republic with energy is the keystone of the Federal Government's policy in the power generation field'.[7] The costs of this commitment dwarfed even those for directly supporting coal (400 million DM in 1977); the Federal government, the mining *Länder*, and consumer levies had had to raise 8.7 billion DM in the period 1974–7. A further commitment of 4.1 billion DM, half of which was to be raised by the consumer levy, was undertaken for 1978. The levy, the coal penny, had initially been proposed by the coal employers and accepted by the interested parties partially as a means of circumventing

EC regulations. It was to fall on private users in such a way that generation costs should be reflected in user charges that would not harm German industry's competitive position. Understandably, the Federal government declared that it 'would use its influence to ensure that the electric power industry will guarantee the use of coal in the long-term and the implementation of the ten-year contract by constructing new power plants'.[8] Electricity generation in the service of the industrial requirements of international performance was to take precedence (and was to apply equally to coal and nuclear energy).

By 1978/9, the proportion of German electricity generated from oil was down to 9 per cent, to be phased out altogether during the course of the 1980s. But more importantly, the Federal government had effected an agreement between the public electricity suppliers and private, industrial electricity producers to supply industrially produced current to the public grid.[9] After the agreement in principle the government set great store on the contracts being drawn up promptly and tabled amendments to the Cartel Act to facilitate their implementation. The West German government had brought about the agreement between two private parties, both of which were organizations representing whole sectors of industry. These powerful, industry-wide organizations could negotiate on behalf of their many constituents and undertake binding commitments. The capacity for industrial self-management, in intimate alliance with public authorities, evinced a continuity of the German industrial tradition which bypassed ideology. In the past, however, this was in order to the pre-empt more extensive state involvement, entailing detailed regulation: pre-emptive co-operation.

The institution of the state and the principle of co-operation continued to characterize the principal functioning of the German energy sector, setting an impressive example for the rest of the economy. But, as has been demonstrated elsewhere in this study, the labour contribution to this system of interest mediation cannot be overlooked. The novelty in this post-war state-concerted self-steering by the associations was now, of course, that the organized representation of employees was also included in the corporative decision-making process.[10]

Pre-War Precedents

Yet, just how new was the involvement of organized labour in this system? A major degree of labour engagement had already been manifest during the immediate aftermath of the First World War:

The integration of the working class into the framework of industrial self-regulation and the system of corporatist interest mediation seemed . . . to be the consequent extension of tendencies of post-liberal interest policy which were already far advanced in some German states, especially in Prussia.[11]

This phenomenon transcended such short-lived arrangements as the Central Working Community of German Employers and Employees (Zentralarbeitsgemeinschaft/ZAG) of November 1918, not least because 'the integration of the working class did not demand a revolutionary change in the mode of German interest policy, but could profit from a strong tradition of collective decision making and co-operation in German business'.[12] This newer development could build upon the corporatist system which had become the dominant feature of economic and social life in Wilhelmine Germany. A similar view is held by Charles S. Maier who maintains that

the key to capitalist continuity in Germany rested upon a parallel behaviour of industry and labour . . . in their very moment of political adversity, major industrialists found it advantageous to secure economic immunities by astute alliance with the trade union-leaders . . . a coalition on the part of key industrialists with organized labour to carve out a corporatist autonomy profitable to both partners.[13]

Industry's need for allies seemed acute at a time when certain members of a social democrat-led government spoke the language of socialization, within a period of post-revolutionary uncertainty. But what precisely was the position of organized labour at this time? Was there also a propensity to co-operate? At the Free Trade Union Congress held on 30 June–5 July 1919 it was proudly claimed that the unions had already forced the employers far down the road to economic democracy.[14] The programme actually adopted in 1919 contained three major emphases: (i) co-determination at every level of the economy; (ii) economic organs of self-government; and (iii) working communities (*Arbeitsgemeinschaften*).[15] Together these were to provide a pattern of labour principles for the remainder of the Weimar period.[16] More specifically, the reconstruction of economic life was to develop in the direction of the commonweal economy, essentially embracing that

complex of ideas which suggested the creation of a set of institutions through which the economy could be controlled and administered in the public interest without interference by the political government . . . an economy planned and regulated by a network of autonomous bodies [*Selbstverwaltungskörper*] composed of employers, employees, the liberal professions, consumers and all other groups of economic significance.[17]

To the many groups who supported economic democracy the programme represented a means of achieving worker participation in industrial and economic decision-making. As discussed in Chapter 7, and as Shuchmann correctly perceived, the ideas which comprised the doctrine of the commonweal economy were originated by Rathenau, developed by Richard von Möllendorf, and propagated among socialist politicians and trade unionists by Rudolf Wissell. But whatever the designation of the socio-economic system and the form that ownership was to take, the unions emphatically pursued the target of worker co-determination at all levels of industry, in concert with the unions, from works level up to industry-wide and national level. The realization of these objectives were to be achieved on the basis of appropriate collective bargaining. Over the ensuing decade the Liberal and Christian unions came to adopt a similar position. The free unions considered economic democracy to be acceptable to all elements of the trade union movement.[18]

Despite the widespread usage of the term 'economic democracy' during the revolutionary period, it was very much a slogan without clearly defined content; it implied the general equality of capital and labour as opposed to the capitalist primacy of capital and the revolutionary socialist's primacy of labour. In the mid-1920s the relevant ideas really began to take on a clearer, more coherent shape, which became evident at the twelfth Trade Union Congress in 1925. From this time onward, the general trend for German labour representation to emphasize the macro-economic co-determination aspect of economic democracy[19] received a distinct boost despite, for example, the apparent failure of the *Zentralarbeitsgemeinschaft*. This macro-perspective in Weimar was most clearly expressed by Naphtali himself.[20] The preoccupation in the second half of the 1920s with the national and district councils led to relative neglect of the works councils none of which were to prove any more effective than the *Zentralarbeitsgemeinschaft* and the organs of economic self-government.[21] Indeed, the district economic councils were never formed.

The German trade union movement could not look back on a long history of supporting active co-operation with employers. It did not formally document a commitment to support of establishment of factory committees (the forerunners of works councils) until its Hamburg conference of 1905. Yet this conference was to prove a watershed in the history of the German labour movement for, first, it implicitly committed the trade unions to the pursuit of an active, harmonious, co-operative relationship with the class enemy, the employer; secondly, it testified to changing power relationships within the social democratic labour movement, marking the coming of age of German trade unions. They asserted their independence from Marxist SPD tutelage by refusing to countenance general strikes (*Massenstreiks*) in the pursuit of political ends, contrary to the position hitherto adopted under SPD guidance. The unions thereby reinforced the reformist-nationalistic elements within the SPD which viewed co-operatives, trade unions, and factory committees as devices that would help reintegrate the proletariat into society. This moderated position was endorsed by the SPD at its Mannheim conference of 1906. And it was an integratist doctrine facilitated by emphasizing points in common between employers and employees/capital and labour, which was the overriding feature of the rise of trade unions to power and influence within the German labour movement.

During the first decade of the twentieth century the basic credo of the trade union movement was 'joint, corporative regulation of labour relations and the establishment of joint arbitration committees';[22] it was the approach unwaveringly pursued by the then head of the Free Trade Unions, Carl Legien. And the implicit stance of official trade union policy before the Great War: 'a kind of (neo-)corporatism within industrial capitalism',[23] became explicit in the 1920s with a clear crystallization of the ideology of economic democracy. However, meaningful co-operation can only occur on the basis of a measure of equality. The demand for equality of labour with capital had been state sanctioned as early as 1890, when Kaiser William II had expressed support for a formal equality and peaceful ('harmonious') relations with responsibility shared by the public power.[24] The basic triadic, corporative structure could not be more clearly delineated, nor the apparent irrelevance of democratic parliamentary procedures.

Curiously a long tradition of support for factory committees and similar arrangements was also evident in right-wing, anti-democratic circles:

The early efforts to establish worker participation in decision-making within the enterprise were rooted in the medieval idea of guilds and corporations, and . . . the discussion of worker participation has never been entirely free of medieval corporatism . . .[25]

A first practical example of participatory arrangements was to be found in the printing industry, one which was operative for approximately forty years. Other employers were to emulate them:

during the second half of the nineteenth century, liberal and paternalistic employers in the Rhineland voluntarily formed such committees that did make contributions to safety and productivity.[26]

The Catholic Church and especially certain Catholic intellectuals such as Franz Hitze also advocated worker participation in company affairs. The primary objective was to overcome class conflict, by promoting harmony between employers and employees—this would be facilitated by developing 'in workers a feeling of belonging to an adequate estate'.[27] However, perhaps the most active advocate of factory councils in the last quarter of the nineteenth century in Germany was a group of conservative reformers, known as 'socialists of the chair' (*Kathedersozialisten*), organized in the Verein für Sozialpolitik (see below and Chapter 7).

The first proposals for macro-co-operation, at the supra-enterprise level, appeared at about the same time, advocated by conservative opponents of the new industrialism:

The essence of these demands has been the establishment and legal recognition of vocational or 'functional' associations consisting of both employers and employees, the organization of councils embracing all persons associated with an enterprise, and the erection of a pyramidal superstructure of local, regional and national economic councils.[28]

But most importantly: 'all the councils are to be empowered to regulate and control various sectors of the economy or the national economy as a whole',[29] thus becoming self-governing organs. The revolutionary ferment of 1848 provided the catalyst for a crystallization of these ideas, a first legislative measure being a Prussian decree of 9 February 1849. The councils thus established may have been short-lived, but the principles behind them lived on to enjoy a renaissance upon the founding of the Weimar Republic.

With the founding of the Second Empire at the beginning of the 1870s a new impetus was given to German industrialization, but the rise of an alienated and politically radical proletariat represented an alarming social question for the established classes and represented a threat to the very stability of German civilization. Two particular socio-intellectual responses took the form, in Schuchmann's words, of 'scholastic' and 'monarchical socialism', which were both anti-parliamentary and anti-*laissez-faire*. The former was particularly influential during the 1870s and incorporated a doctrine which advocated worker participation in decision-making on every level of the economy, but especially on the supra-enterprise level. Such views exercised a particularly strong influence within German Catholicism. Yet by the end of the century the desire of scholastic socialists to reconstruct society had given way to that of melioration, a development representing convergence with that other, potent school of thought, the monarchical socialism of the 'socialists of the chair'.

The intention was to make the monarchy particularly active in political and economic reform, a primary aim being to defeat Marxian socialism. It was argued

that class antagonism could be overcome only through the 'development of an organized, corporative system for reaching agreement between employers and workers' and through the establishment of a 'complete scheme of representative associations for both parties . . .' Such institutions, enabling workers to participate in decision-making, it was believed, would lead to an awareness of the essential harmony of interests of labour and management, to mutual respect and a resulting attenuation of social conflict.[30]

Such views found wide currency beyond the circle of initial propagators, their most eminent practitioner being Bismarck himself, who, with his successors under William II, pursued the doctrine of a 'social monarchy of the Hohenzollerns'.[31]

Although Bismarck was thwarted in his ultimate objectives, 'the idea did not die, and in the period directly after World War I, the Bismarckian plans found many advocates among political leaders'.[32] Indeed, even after the outbreak of war the idea of worker participation at this supra-enterprise level had, remarkably, appealed to many diverse groups: 'It appealed to Protestants as well as to Catholics, to Social Democrats as well as to Conservatives, to authoritarians as well as to democrats.'[33] In pre-war imperial Germany the non-democratic tradition seemed to be far stronger in the respective fields of political theory and practice than socialist alternatives. This tradition advocated 'a new form of social organization, the fundamental unit of which would be an occupational group rather than an individual or economic class'.[34]

The foundations had certainly been well laid for the creation of that consensus indispensable to a viable, functioning corporatist system. 'In the long run the major possibility for social consensus derived from a slow transformation of the principles of class division'.[35] This process was encouraged at the very outset of the 1920s by influential sections of the major party of the left, the SPD, who ceased to perpetuate class warfare; 'working class power . . . yielded to a class-neutral technocratic concept',[36] within the context of support for the commonweal economy, a central pillar of economic democracy. Such a development is perhaps not surprising when one considers the common ground which could exist already in the early 1920s between left and right radicals. It was particularly apparent regarding certain political demands to organize society on the basis of occupation,[37] which seemed for a time to attract considerable support both among top industrialists and the Independent Socialists (USPD).

But the philosophy and associated structures of a fully organized corporatism, despite the practical precedent of coal industry reorganization, did not marry well with the Anglo-Saxon model of liberal parliamentary democracy. A transitional period of adjustment was required which had not been available before the war and was certainly not possible afterwards. The economic strains imposed upon this untried system during the 1920s proved too great. Full parliamentary democracy became identified with economic chaos, political failure, and with social disintegration, as feared previously by its liberal opponents. Nazism restored order and stability of an extreme yet temporary kind. After the catastrophe of the Second World War, a practical compromise of the two systems had to succeed in order to avoid another disastrous regression and to pre-empt developments leading to the left totalitarianism of Communism.

Two major middle ways were to emerge, with ideologies to sustain a politico-economic practice and which, despite the arguments of purists on both sides, were not mutually exclusive. The social market economy and economic democracy possessed more points of contact than of opposition, they were flexible theoretical constructs, not irreconcilable, rigid bodies of doctrine. Both these philosophies represented an ideological pragmatism well suited to a nation requiring political, social, and economic reconstruction. They offered a theoretical coherence and consistency which prevented urgent, practical needs of policy degenerating completely into a series of uncoordinated, *ad hoc* measures without system or perspective. In short, a rationale was provided which assisted objectification of policy-making.

Labour's contribution of economic democracy, which was initially, in the 1920s, a synthesis of a proletarian and a bourgeois ideology, came to shed most vestiges of any Marxist origin. The formulation which Schuchmann applies to 'full co-determination' applies equally to economic democracy:

[it] defines an economic organization which is related to corporatism of a non-fascist kind, as well as to syndicalism, particularly its Guild Socialist variant ... full co-determination implies a corporative economic organization within a democratic and pluralistic political and social organization.[38]

This full co-determination was described by Golob as 'democratic corporatism', which is also a precise description of economic democracy. This is German labour's active theoretical and practical contribution to German neo-corporatism. German labour has consciously and positively opted for a co-operative, compromise stance as opposed to a conflictual ideology. In 1968 the DGB expressed this view in the following terms:

Co-determination... should and can lead to tension and conflict being resolved more fairly. Co-determination guarantees that the resolution of conflict is to take place within certain institutions according to certain rules whereby all participants strive for solutions which do justice to the enterprise as a whole and to the interests present within it. The German trade unions consider the fair solution to conflict according to rules binding on all participants, within the institutions of co-determination, to be better.[39]

Within economic democracy corporatist arrangements are explicit; within the practice of the social market economy they are implicit.

The Trade Unions and Economic Democracy

There can be little doubt that the claim by the German labour movement to have followed, after the Second World War, the path marked out by Fritz Naphtali during the later years of the Weimar Republic is broadly justified. It is a view supported by much evidence and therefore supported by many shades of academic opinion including neo-Marxists. In a number of resolutions at certain trade union conferences it became clear that the heritage of economic democracy was very real and strong. At the very first trade union conference of the British zone of occupation on 12 March 1946 a resolution calling for 'plant- and extra-plant level co-determination' was considered and in the same zone, at Bielefeld on 21–3 August 1946, democratization of the economy was demanded, with works-level co-determination as an immediate first step along the path to socialization. Similarly the second interzonal conference of German trade unions held at Hanover on 18–19 December 1946 passed a resolution demanding reconstruction of the German economy on a democratic basis, with effective, direct trade union and works council involvement.

These resolutions on democratization and co-determination were by no means the only ones of their kind, nor were they the last. But they did begin to demonstrate two important things: first, that the home and powerhouse of economic democracy and co-determination after the war was still the German trade union movement; and secondly that the philosophy in question was filling the ideological gap left by the SPD between the wars. Despite the Marxist terminology, as seen in the Heidelberg Programme of 1925, the SPD had ceased in the 1920s to offer a revolutionary alternative to capitalism. Along with the rest of German labour organizations generally, it 'had become a liberal democratic movement, concerned with a capitalist welfare state, not a radical transformation of society'.[40] By the end of the 1940s the West German trade unions were well on the way to providing a similar ideological stimulus to the SPD just as they had after the First World War.

In February 1947, the CDU promulgated its famous Ahlen Programme. It demanded, among other things, that necessary economic planning and steering (*Lenkung*) be undertaken by industrial self-governing bodies in which employees and consumers had equal representation along with the employers. Three months later a resolution was carried during the fourth interzonal trade union conference in Garmisch Partenkirchen which demanded:

> Erection of a Central German Office for Economic Planning and Direction and construction of a system of organs of economic self-government. On these bodies, as well as in the control of the Central Office, the trades unions must have completely equal representation.[41]

The *Land* Government Declaration of North-Rhine Westphalia one month later (17 June 1947) also advocated economic self-government and business chambers with parity representation of capital and labour.

But the 1949 general election represented a major defeat for the aspirations of the German labour movement. Organized labour responded by uniting its industrial unions in a federal organization, the Deutscher Gewerkschaftsbund (DGB), and by promulgating in Munich a first coherent programme of reform, pursuing demands shaped by the mould established under Weimar. As far as co-determination and the economic order were concerned the 1949 Munich Conference declared, in Potthof's words, that:

> the ideal picture of the economy entertained by the trade unions was a commonweal system . . . a mixture of market economy and planned economy . . . Private ownership of certain key industries was to be transferred to common ownership. The organs of economic self-government and of works management, especially in large works, was to be democratized by including representatives of the workers.[42]

Within less than a year organized labour became embedded in a bitter defence of parity co-determination in the steel industry, involving the whole *montan*-sector. Labour proposals for reorganization of the economy in general and its demands for full co-determination in particular appeared to enhance confrontation during that heated debate, but the real, basic position of the unions was one of promoting harmony rather than conflict. This was confirmed by Hansler, a leading SPD spokesman in the Bundestag, who said of Hans Böckler's role, and that of the trade unions, that 'He strove for a platform for a broadened and deepened co-operation of the social partners'.[43] Such was the objective which was not to be obscured by controversy over specific industrial disputes.

Thus although organized labour may generally have approached relations with the employers in a spirit of compromise, and, indeed, specific legislation, such as the Shares Law (*Aktiengesetz*) and the Works Council Act of 1952, may have required co-operation on the supervisory boards, in themselves they were no guarantee that in particular industries harmonious relationships would prevail. In this context, the role of the labour director in the *montan*-sector was later to force itself upon the attention of the authors of the Zwischenbilanz study,[44] indicating again and again its problematic nature and importance. In the mid-1950s Schuchmann had asserted that 'one of the major tasks of the labour director was to foster co-operation between the firm's personnel and its management',[45] and he had also predicted that 'the ultimate success or failure of co-determination within an enterprise depended largely on the ability of the labour director to foster among the workers the feeling that they are partners in a common enterprise.'[46]

Neuloh was to establish that the primary function of the labour director 'as an office of arbitration or peacemaking' was as follows:[47] he could restrict or resolve conflicts at works level, without them becoming a company- or even an industry-wide dispute. Given this key role, it is significant that although the *montan*-labour director could not be elected against the wishes of the labour representatives on the supervisory board, there is also no evidence to suggest that he was ever elected against the wishes of management. The Biedenkopf Commission Report was to confirm that the position of the labour director reflected the integrative role of co-determination.

Trade union sponsored research in the early 1960s appeared to confirm that the resolution of industrial conflict, by means of co-determination, both enhanced and reflected conflict resolution in wider society. Neuloh referred to a dual power structure in the firm, reflecting a dual power structure in society, one where class conflict was not so much institutionalized but transcended and overcome. The Zwischenbilanz study was able to conclude that antagonism between employer and worker had become, through co-determination, transformed more into co-operation so that 'antagonism has become more like togetherness'.[48]

The first coal crisis appeared not to subject employer–labour relations to insurmountable strain, especially when the focus was on the board of management. Although it would seem that the position of the labour director was weaker on those boards of management containing a total of more than three directors, the other directors seem not to have exploited this situation for the number of divisive votes (*Kampfabstimmungen*) did not increase during the coal crisis, often indeed discussion was prolonged until a unanimous view was established. On the other side, even in companies where the influence of the labour director was particularly strong, employees tended not to exploit the company's vulnerability: 'the will of the employee side was never forced upon the company, even where this would have been possible.'[49]

The spirit of compromise appeared to be highly developed on both sides. Overall, the various conflicting pressures tended to balance themselves out during the course of the crisis and general relationships did not deteriorate as seriously as was frequently believed outside the industry. The main enduring result of the heightened co-determination activity during the crisis, in terms of practical innovation, was the new social plans which represented a great success for plant-level co-determination. This micro-corporatism was also paralleled at a macro-level. During the first coal crisis the demands for common ownership were renewed by IG Bergbau, but they had to take second place to advocacy of a more active and comprehensive scheme for involving all affected parties in managing the industry's crisis; this entailed proposals for an Energy Council and a Coal Council.

A New Direction for Organized Labour?

In November 1963 the DGB presented its new programme. Although it could be criticized as being no longer connected with economic democracy and a 'Third or Middle Way' nevertheless in the Preamble it affirmed that:

The trade unions fight for the extension of employee co-determination. With this they want to inaugurate a transformation of the economy and society with the objective of letting all citizens participate in economic, cultural and political decision-making.[50]

This broad social perspective was stressed in H. O. Vetter's assessment of the programme's 'strong emphasis on the idea of co-determination and on the dedication associated with it concerning the democratization of the economy and society'.[51] Although the term 'economic democracy' was nowhere employed in the programme its preoccupation with comprehensive democratization was very much in the tradition of the central thrust of economic democracy.

The document continued to take co-determination within the economy, especially parity co-determination, very seriously. For industry as a whole the *montan*-solution was required, and the business chambers were also to evince equal capital and labour representation. The programme contained no references to common ownership and few to the commonweal economy; presumably the trade union position was like that of the SPD as demonstrated in its Godesberg Programme: 'commonweal economy solutions' were to be employed as a last resort, on the usual basis of self-government. Neither of the two programmes referred to nationalization or socialization. The SPD viewed *montan* co-determination as the beginning of a 'new ordering of the economy' and, like the latest trade union programme, envisaged extending it to major companies.

Changed external societal circumstances decreed that the DGB's new programme would significantly differ from the previous one, both in mode of presentation and in the choice of means to implement objectives. Indeed, circumstances are changing continually and the environment at the outset of the 1970s was as different again from that at that beginning of the 1960s. In April, and again in November, 1971, the DGB published a major tract: 'Co-determination—A Demand of Our Time'. Its primary purpose was a restatement, in the changed environment of a social democratic–liberal ruling coalition, of the principles and intentions of the Basic Programme of eight years before. It claimed a very broad remit for trade union activity, representing all employees, whether organized or not. The document states categorically that the trade unions are the main inspiration behind labour's co-determination demands. It asserts that the extension of *montan* co-determination was fully justified in view of its success, particularly in the Ruhr where it had 'contributed to a solution of a structural crisis and avoided serious political and social tension'.[52] Indeed, the document ended with an assertion of this central belief that co-determination is a rational means of dealing with conflict, not least in preventing political radicalization.

The theme of conflict resolution and the positive, constructive contribution of organized labour was taken up by Vetter and remained a recurrent topic in many of his speeches throughout the 1970s. He presented the case forcefully and simply: 'Co-determination is and remains the splendid offer of employees to society and the economy to settle conflicts in a progressive manner.'[53] That was the heart of the matter, defining the essence of labour's contribution to the German corporatist system. It explains the significance of his earlier claim that German trade unions had never been pure interest groups and that 'they have always acted co-operatively and responsibly in shaping state and society'.[54] That this does not simply represent special pleading is confirmed by Thimm:

> the structural changes in the *montan*-industry during the 1950s and 1960s could not have been carried out without the full support of the national union leadership and their representatives on the supervisory boards, who placed long-run regional and national interests above the short-run concerns of employees and shareholders.[55]

The rewards of such co-operation were by no means insignificant: in the *montan*-sector labour had acquired a *de facto* veto over all board of management appointments, not just those of the labour director.[56]

In the 1970s, democratization was still very much the final objective, to be realized by a maximum amount of co-determination at all levels simultaneously. The debt to Naphtali is apparent, as was repeatedly acknowledged by Vetter himself. Half a century later, in the second half of the 1970s, a major redefinition of economic democracy was attempted by Fritz Vilmar, a trade union academic who had become a professor of politics at the Free University of Berlin. Vilmar took pains to stress that economic democracy is still 'a Third Way',[57] openly acknowledging the historical debt going back to the 1920s. He clearly returns the argument to one of 'self-government', in this case 'workers' self-government' (*Arbeiterselbstverwaltung*). For Vilmar, the highest goal of economic democracy is the 'greatest possible overcoming of the alienation of working people'.[58] Logically, the primary means to achieve this is self-governing plants, for the ultimate objective can alternatively be expressed as the 'self-determination and self-government of working people'. Co-determination represents an essential intermediate stage along this road. The most appropriate designation of the necessary interaction between employees and employers is not, according to Vilmar, a misplaced form of social 'partnership' but 'antagonistic co-operation', an organizational principle embracing two allegedly complementary alternatives, ultimately to be transmuted into 'optimal self-government'.

Labour, Co-determination, and Social Harmony

However, whatever the theoretical desirability of fundamental reform, it could not be denied that harmony had prevailed since the early 1950s. On the employer side there was no friction between management boards and the owners represented on the supervisory boards.[59] In the relationships between works councils and the trade unions in the period 1952–72 real and open conflicts rarely occurred; in fact, works councils had proved to be 'an effective device for employee participation in the organization of their workplace during the period 1952–72; they improved the security of individuals and protected them from arbitrary hierarchical decisions'.[60] And in the ensuing eight years Thimm had to conclude that 'operationally nothing has changed'.[61] Similar concord appeared to exist in the non-*montan*-sectors on supervisory boards: 'top managers from all sectors have repeatedly praised the loyalty and entrepreneurial attitude of the internal employee board members.'[62] Co-determination in industry represented the institutionalization of that 'traditional German principle of social co-operation', a form of informal arbitration contrasting with, yet complementing, formal arbitration procedures:

> The willingness to arbitrate social conflicts epitomized the principle of social obligation and co-operation that has prevailed under the Weimar Republic, National Socialism, and the current Federal Republic.[63]

Enduring values and attitudes are prior to their formal institutionalization, not vice versa.

If the relatively harmonious success of co-determination at works and company level were to be extended to the supra-enterprise sphere, then this would begin fundamentally to alter the role, *raison d'être*, and self-perception of the trade unions themselves, and for this they are well prepared. The union leadership realized that equal participation could not be obtained without accepting equal responsibility; the unions would, therefore, be transformed into a different institution whose function would no longer be confined to wage bargaining.

Union ideologists such as Dr Hans-Adam Pfromm, IG Metall's expert on co-determination, have emphasized the broad social responsibilities of the unions; they have developed from mere advocates of labour interests into the social conscience of the nation. The loss of *Tarifautonomie*, according to Pfromm, is no longer important in a society in which capital and labour participate equally in economic decision-making. The unions have outgrown their role as worker advocates.[64] Whatever the antipathy of some West German trade unionists towards notions of social partnership, it would seem that what could be described as organized labour's corporatist integration is virtually complete.

The Social Market Economy and Economic Democracy

Despite the continuing commitment of the West German labour movement to the ideology of economic democracy, in practice it experienced very little difficulty in reconciling itself to the prevailing ideology of the social market economy. For an explanation of the compatibility of these ideologies it is necessary to go beyond superficial notions of mere institutional opportunism or working-class betrayal. There is an important area of common ideological ground which has been consistently pursued by German labour and which centres on the principles of the market, competition, and the necessary decentralization of decision-making associated with the effectiveness of both. With the passage of time the implicit dedication of both wings of the labour movement, industrial and political, to these three principles became more explicit.

An early trade union voice was that of Georg Berger of the mining union, in his plans for the socialization of the coal industry. His proposals were intended to establish 'a healthy competition between the pits'; this approach reflected the intention to restore to the price mechanism its directional power. Later the same year in the *Land* of Hessen, a *Land* in the vanguard of social democratic aspirations for creating a Third Way, the founding of industrial social communities was advocated by the SPD. These commonweal enterprises were very much intended to face the rigours of free competition. A cardinal feature of the new economic system would be the 'principle of economic decentralization'.

By the mid-1950s the new Federal Republic's social market system appeared to be consolidating fast and the SPD itself began to emphasize its commitment to a market-type economy and appropriate decentralization. The apotheosis of the trend was confirmed at the party's famous 1959 Bad Godesberg Conference, not least in the section of the relevant document calling for public control of economic power, one means in particular being common ownership organized according to the principles of decentralization.

One year earlier, in 1958, the head of IG Bergbau, Heinrich Gutermuth, even at the height of the first coal crisis, was to call for real competition and genuine decentralization, whilst still advocating common ownership. On behalf of his union he proclaimed: 'We want to put only feigned competition really into practice.'[65] His recommendations would, apparently, not only enhance healthy competition, but also further the principle of decentralization. With both the SPD and a major union appearing to be very open to what had traditionally been considered capitalist principles, it was perhaps only a matter of time before the rest of the labour movement would also formally accommodate itself to such developments.

Over the next decade the trade union movement came to accept not only the profit principle associated with the private market system, but it came also to realize more clearly that its very own main demand for company co-determination was, in fact, dependent on the existence of a decentralized economic system. This theme of the compatibility of co-determination and market economy, indeed their mutual dependence, became a constant one in the speeches of the head of the DGB, H. O. Vetter. Throughout the 1970s he affirmed that co-determination was not incompatible with competition and the pursuit of profit. Moreover, the support of labour representatives within the *montan*-industry's co-determination institutions for these principles had been confirmed by the Biedenkopf Commission, the former had never even questioned their validity.

Vetter stressed that employees and their representatives knew that the main prerequisite for the maintenance of employment in a market system was the profitability of a company. In a sense, the worker was now the true capitalist for the natural long-term interests of employees ensured observation of the profit principle in company policy. It would appear that, in the official trade union view, full co-determination would not endanger the prevailing social and economic system; on the contrary, without co-determination it would not overcome the serious challenges to which it was then subject.[66]

It was not only the practitioners of trade union affairs, the leading officials, who recognized the necessity for functioning markets; it was equally evident among theorists such as Vilmar who asserted that the theory of economic democracy recognizes the useful functions of certain regulating mechanisms of the market. Vilmar was particularly concerned by the power of trusts to distort and control markets and, like Müller-Armack, he envisaged the state playing an active role in promoting real competition. His micro-economic view of the market's function was in no respect anything fundamentally new, it was very much in the mainstream of social democratic thought over at least the preceding decade.

It is therefore apparent that the social democratic labour movement in the Federal Republic increasingly embraced apparently capitalist beliefs both regarding competition and decentralization which had long been accepted by its leading activists and regarding markets and profits which followed both logically and practically in the changing external environment. The inherent points of contact between economic democracy and the social market economy were significant, and, if one substituted the one term for the other in the following quotation from Ludwig Erhard, it would seem to make very little material difference:

The 'social market economy' developed as the remainder of a 'revolutionary synthesis' of the 'organically grown and proven elements of the market economy' and the 'system of economic planning and steering'.[67]

There thus seemed to be two middle ways which, within one generation, were to demonstrate considerable convergence. Although differences of principle and intention existed, the systems were by no means incompatible and not mutually exclusive.

The association, if not integration, of apparently opposing philosophies in relation to the practical organization of the Federal German political economy was further facilitated by a phenomenon variously referred to as the dichotomy between ownership and control, or alternatively as the managerial revolution. Important, early discussions of some of the issues involved in post-war Germany took place in the North-Rhine Westphalian *Landtag* in those first debates on the proposed socialization of the coal industry.

Not that managerial control had not been unimportant before the war, either in Germany or in other industrial nations, but, as Schuchmann rightly points out, certain special factors obtained in Germany: first, there was the Joint-Stock Company Act (*Aktiengesetz*) of January 1937, which was still the operative company law until 1965 (nor was it fundamentally revised even then) which 'deprived the shareholders of virtually all their rights of use and disposition';[68] secondly, there was the control over companies exercised by cartels and industrial associations, both tending to be dominated by large concerns.[69] Despite public controls on formal cartels in the Federal Republic, informal cartels and powerful industrial associations ensured the continuity of this tradition. Thirdly, there was the credit mechanism: 'Although legally only shareholders, individual proprietors, and partners are regarded as the owners of an enterprise, from an economic point of view creditors are also owners.'[70] There is little evidence to suggest that this situation, which accentuated managerial pre-eminence, fundamentally changed in the Federal Republic. On the contrary, the centrality of the banks as intermediaries, or even as economic owners, appears to have become more firmly established than ever.

Trade unions had not been the only organizations desiring a complete restructuring of society and the economy in the immediate post-war period and for many interests the issue of property ownership was secondary. As far as the particular issue of parity co-determination in the *montan*-industry was concerned 'it was always only a matter of the most appropriate means of participating in the control over property'.[71] Such an approach was tenable for technical reasons because 'The position of the employer in labour law is not tied to particular ownership relations'.[72] Co-determination was aimed at control rather than at property as such: 'Co-determination is not directed against property rights, which are guaranteed by ownership of the individual share, but only against the control rights of a share.'[73]

But it was precisely this attempt to influence control in company affairs to which private employers most strongly objected, viewing co-determination as 'only a substitute for socializing the economy: the socialization of property is renounced in favour of socializing control over property'.[74] Control would be exercised where company policy was formulated, apparently on the supervisory board, a body, according to Neuloh, acquiring growing importance, and one which demonstrated relative independence from shareholder influence. Not that diminution of shareholder influence necessarily implied its replacement by a negative worker influence in *montan*-companies; on the contrary, it was precisely the employees who had proved so keen on investment to increase company competitiveness by modernization. But employers were generally very hostile to the whole principle of parity co-determination and no amount of positive indicators from the world of industrial practice would modify that attitude.

In the meantime, it had become customary for the trade union movement to stress co-determination at company level rather than at any other and to focus primarily on the supervisory board. But by the 1970s research was indicating that the board of management could often be considered the real fulcrum of company power in terms of decision-taking. Thimm highlighted two particular instances of the considerable, autonomous influence of the board of management in contemporary business affairs: first, 'Although segments of the German steel industry are owned by provincial and Federal governments, there has been no difference in enterprise policy between private and nationalized firms'.[75] This is partly because of the 'market orientation of German managers', but also because of the legal position of the board of management which is relatively insulated against pressure of owners, whether public or private. Secondly, the serious steel strike of 1978, in an industry with full *montan*-co-determination, indicated how independently boards of management could act. They were the managers represented by the forthright employers' associations and were regarded as employers against whom the strike was directed. Labour influence on the supervisory boards and on the board of management itself was not sufficient to dictate terms to management acting as the employer.

Nevertheless, whatever the problematic nature of relations between the board of management and the supervisory board, for the German labour movement at the end of the 1970s the distinction between control and ownership was still of fundamental importance and likely to remain so for the foreseeable future. Labour's concern was with control as an end rather than with property as a means. Economic democracy conceived of common ownership as one option among several and the social market economy clearly did not preclude such ownership. Practical decisions on the particular form of the 'ownership of the means of production' would be decided by value judgements, influenced as much by general socio-cultural considerations as by 'matters of fact' regarding technical efficiency. For the mining union such issues were academic, in the words of its chairman Adolf Schmidt:

> The realization of the comprehensive trade union concept of 'economic democracy' was, in essence, achieved in mining with the establishment of *montan* co-determination.[76]

Yet, despite inevitable differences of opinion between different unions and within individual union hierarchies, the West German trade union movement as a whole displays on overall unity and coherence which remains vital to its strength and influence. Such inner concertation, assisted by a shared ideology, is a necessary condition of trade union effectiveness within the wider system of social corporatism. Theoretically, co-determination is both aggressive and conciliatory, aggressive because it wishes to put an equal share of industrial power into labour hands, which means taking extensive powers away from capital, conciliatory because it is based on an equal partnership and not on subordinating capital to labour. In practice, the empirical evidence that the institution of co-determination both strives for, and largely achieves, industrial harmony is irrefutable. If it is the case that a primary function of the German state is actively to facilitate the conciliation of conflicting interests, then the state will be obliged to further the efficacy of the institution of co-determination. But conciliation also requires appropriate behaviour from employers—paradoxically, although the evolution of German co-determination owes much to a long tradition of progressive employer activity, in the central *montan*-sector such initiatives were conspicuous by their absence, not through indifference but through distinct hostility. Yet at an early stage in post-war German industrial reconstruction there was real evidence of fundamental change in the attitude of coal employers, both before and after the implementation of parity co-determination. In Kost's words:

> that during the whole time of the existence of the DKBL there was not a single instance where, even if on some occasions only after intense discussions, unanimity

had not been achieved in the views of both groups on the advisory council and between the advisory council and the directorate.[77]

The implications are clear: if healthy industrial relations could be, and were, established in what had traditionally been the most difficult of Germany's major industries, then that would augur well for industrial relations generally in the Federal Republic. By anchoring parity co-determination in the problematic *montan*-sector, a pre-emptive strike had been undertaken in favour of forcing co-operation between employers and employees. Such a major step had been instrumental in opening up the existing corporatist system, facilitating the acceptance of labour into a more genuinely triadic structure.

In the 1980s a formidable constellation of factors threatened the position of organized labour and hence threatened to undermine this very system. In particular the 1980s saw a Conservative CDU-led government coalition; the Neue Heimat and Co-op Bank fiascos;[78] a more aggressive stance taken by certain employers, especially in steel; and continuing high levels of unemployment. But despite these challenges organized labour retained its strength and resistance. Markovits's conclusion was as valid as ever, 'Ultimately, the heavy burden of Germany's past will remain the most reliable guarantor of stable capital–labour relations in the Republic'.[79] The unification of the two Germanies is more likely to reinforce than to undermine that truism.

Notes

1. W. Abelshauser, 'The First Post-Liberal Nation: Stages in the Development of Modern Coporatism in Germany', *European History Quarterly*, 14 (1984), 286.
2. See Z. Jakli, 'Staatliche Intervention und private Politik im Energiesektor', in V. Ronge (ed.), *Am Staat Verbei* (Frankfurt: Campus Verlag, 1980), 53.
3. Ibid. 58; and the 'financial basis and implementation had to be created and ensured by the state' (*loc. cit.*).
4. By no means a new phenomenon: cf. Charles S. Maier, on the period following the First World War: 'in Germany self-interest suggested the formation of new pressure groups and collective accommodation with other social forces' (C. S. Maier, *Re-casting Bourgeois Europe* (Princeton, NJ: Princeton University Press, 1975), 81).
5. Jakli, 'Staatliche Intervention', 60.
6. Federal Ministry of Economics, *Energy Policy Programme for the Federal Republic: Second Revision* (14 Dec. 1977), 5.

7. Ibid. 20
8. Ibid. 21.
9. H. Schmidt, 'Erklärung der Bundesregierung zur Energiepolitik nach dem Europäischen Rat und dem Weltwirtschaftsgipfel', *Bulletin*, 87 (July 1979), 814.
10. W. Abelshauser, 'Von der Kohlenkrise zur Gründung der Ruhrkohle AG', in U. Borsdorf and H. Mommsen (eds.), *Glück auf, Kameraden! Die Bergarbeiter und ihre Organisationen in Deutschland* (Cologne: Bund-Verlag, 1979), 443.
11. Abelshauser, 'The First Post-Liberal Nation', 298.
12. Ibid.
13. Maier, *Re-casting Bourgeois Europe*, 54.
14. See B. Otto, *Gewerkschaftsbewegung in Deutschland* (Cologne: Bund-Verlag, 1975), 70.
15. Compared with 'The major proposals of the socialist left during 1919–1920 centred around factory councils, nationalization and economic planning' (Maier, *Re-casting Bourgeois Europe*, 138).
16. A. Shuchmann, *Co-determination: Labour's Middle Way in Germany* (Washington, DC: Public Affairs Press, 1957), 66.
17. Ibid. 32.
18. If not necessarily as a means to socialism—and a fuller development of worker participation had also become a standard component of the programmes of all parties of the centre and left.
19. Cf. E. Potthoff, *Der Kampf um die Montanmitbestimmung* (Cologne: Bund-Verlag, 1957), 11 and 15.
20. 'The constitutional and psychological pre-requisites for an era of large-scale self-government, i.e. for a far-reaching transfer of state functions to economic associations, had been created.' Quoted in F. Deppe, *Kritik der Mitbestimmung* (Frankfurt: Suhrkamp Verlag, 1969), 50.
21. For a less negative view of the success of the works councils, see A. L. Thimm, *The False Promise of Co-determination* (Lexington, Mass.: Lexington Books, 1980), 6–7.
22. A. Klönne, *Die deutsche Arbeiterbewegung* (Düsseldorf: Eugen Diederichs Verlag, 1980), 112.
23. Ibid.
24. See the Imperial Decree of 4 Feb. 1890, reprinted in K. Buchholz *et al.*, *Mitbestimmung und Interessenvertretung* (Cologne: Bund-Verlag, 1981), 21.
25. Shuchmann, *Labour's Middle Way*, 14–15. But the distinction between participation at works/factory level and at company level was by no means always clearly drawn.
26. Thimm, *False Promise of Co-determination*, 4.
27. Shuchmann, *Labour's Middle Way*, 16.
28. Ibid. 23.
29. Ibid. 25.
30. Ibid. 33 and 34.

31. He intended replacing the Reichstag, by a *coup d'état* if necessary (see H. Heffter, *Die deutsche Selbstverwaltung im neunzehnten Jahrhundert* (Stuttgart: K. F. Koehler Verlag, 2nd edn., 1950), 687), with a corporative chamber representing vocational interests (not least because of its potential for manipulation, 'ihre politische Fügsamkeit', ibid.), initially as an advisory chamber. However, he only succeeded in establishing an advisory council for the Prussian *Landtag*.
32. Shuchmann, *Labour's Middle Way*, 36.
33. Ibid. 40.
34. Ibid. 39.
35. Maier, *Re-casting Bourgeois Europe*, 584.
36. Ibid. 143.
37. See especially F. Glum, *Selbstverwaltung der Wirtschaft* (Berlin: Hermann Sack Verlag, 1925), 136 and 137.
38. Shuchmann, *Labour's Middle Way*, 236.
39. From the 1968 DGB publication 'Employee Co-determination: A Guide', repr. in F. Vilmar, *Politik und Mitbestimmung: Kritische Zwischenbilanz— Integrales Konzept* (Frankfurt: Athenäum Verlag, 1977), 36.
40. Shuchmann, *Labour's Middle Way*, 56.
41. R. Judith *et al.*, *Montanmitbestimmung: Dokumente ihrer Entstehung* (Cologne: Bund-Verlag, 1979), 94.
42. Potthoff, *Kampf um die Montanmitbestimmung*, 140–1.
43. Quoted in Deppe, *Kritik der Mitbestimmung*, 104.
44. O. Blume, H. Duvernell, and E. Potthoff, *Zwischenbilanz der Mitbestimmung* (Tübingen: J. C. B. Mohr (Paul Siebeck), 1962).
45. Shuchmann, *Labour's Middle Way*, 152.
46. Ibid. 154.
47. O. Neuloh, *Der neue Betriebsstil* (Tübingen: J. C. B. Mohr (Paul Siebeck), 1960), 164–5.
48. Blume, Duvernell, and Potthoff, *Zwischenbilanz*, 220.
49. C. D. Schmidt, 'Die Krise im Steinkohlenbergbau und ihre soziale Problematik unter besonderer Berücksichtigung des Ruhrgebietes', Ph.D. thesis, Westfälische Wilhelms-Universität, Münster, 1967, 396.
50. Quoted in H. C. F. Liesegang (ed.), *Gewerkschaften in der Bundesrepublik Deutschland* (Berlin: Walter de Gruyter, 1975), 43.
51. Quoted ibid. 56.
52. See S. Hergt (ed.), *Mitbestimmung* (Opladen: Heggen-Verlag, 2nd edn., 1974), 124.
53. H. O. Vetter, *Mitbestimmung: Idee, Wege, Ziele* (Cologne: Bund-Verlag, 1979), 58.
54. Ibid. 14.
55. Thimm, *False Promise of Co-determination*, 47.
56. That co-determination equals co-operation was confirmed by 'the German

Minister of Labour, Walter Arendt (SPD), who greeted the passage of the 1976 law by saying that he anticipated the beginning of a "new era of co-operation, without parallel in history" ' (ibid. 86).

57. F. Vilmar, 'Mitbestimmung und Humanisierung der Arbeit', in id., *Theorie der Wirtschaftsdemokratie*, Course 1 of a series of lectures for the West German Open University (Fernuniversität), Hagen, 1980, 21.
58. Ibid. 29.
59. D. Brinkmann-Herz, *Die Unternehmensmitbestimmung in der Bundesrepublik* (Cologne: Verlag Kiepenhauer & Witsche, 1975), 59.
60. Ibid. 46.
61. Thimm, *False Promise of Co-determination*, 96.
62. Ibid. 50; see also 56.
63. Ibid. 7.
64. Ibid. 120.
65. M. Martiny and H. J. Schneider (eds.), *Deutsche Energiepolitik seit 1945* (Cologne: Bund-Verlag, 1981), 132.
66. The recession induced by the oil crisis and associated with the high unemployment of the mid- to late 1970s.
67. Quoted in Otto, *Gewerkschaftsbewegung in Deutschland*, 92.
68. Shuchmann, *Labour's Middle Way*, 165.
69. '... which may be said to epitomize control by non-owners' (ibid. 185).
70. Ibid. 186. He continues: 'As a result, a large proportion of the means of production in a modern economic organization "belongs" to one group of people and is controlled by another group.'
71. Potthoff, *Kampf um die Montanmitbestimmung*, 78.
72. Ibid. 121.
73. Ibid. 144.
74. Deppe, *Kritik der Mitbestimmung*, 179.
75. Thimm, *False Promise of Co-determination*, 77.
76. See the foreword by Adolf Schmidt to August Schmidt, *Lang war der Weg* (Bochum: IGBE, 2nd edn., 1978).
77. H. Kost, 'Die Tätigkeit der Deutschen Kohlenbergbauleitung-Schlußbericht', *Glückauf*, 90 (1954), 91.
78. Both controlled by the DGB-owned holding company BGAG (Beteiligungsgesellschaft für Gemeinwirtschaft AG). Cf. article 'Durchgriffshaftung auf den DGB und die Holding der Gemeinwirtschaft gefordert', in *Handelsblatt*, 184 (1980).
79. A. S. Markovits, *The Politics of the West German Trade Unions* (Cambridge: Cambridge University Press, 1986), 449.

10

Conclusion

Coal and Contemporary Organized Capitalism

The West German political economy may very well evince a framework with characteristics that can readily be described as neo-corporatist; the West German labour movement may very well be committed to a constructive, conciliatory ideology; it seems also to be an empirically well-established fact that the institution of co-determination has functioned successfully in the West German coal industry. But is there a common element to all three phenomena which, both as concept and institution, links and combines them in such a way as not only to provide continuity and coherence in their association and evolution but also associates them in an intimate way with that enduring power phenomenon in Germany, the state? The present study as a whole and Part III in particular suggests that a German tradition exists, that of industrial self-government, which facilitates concertation, and often integration, of interests regarding industrial policy inputs and continuity in industrial policy output. This phenomenon is strong and flexible enough to manifest itself in various forms in different industries and at different times while remaining robust and resilient enough to ensure that the necessary co-ordination of effort required to maintain Germany's premier position as an industrial nation is not impaired.

Industrial Self-Government the Key?

Not only is the concept of industrial self-government common to the neo-Marxist theory of organized capitalism, to the labour ideology of economic democracy, and to neo-corporatist political analysis, it would also appear, on the basis of the present study, that it is THE institution that characterizes the organization and functioning of the German political economy. This institution of self-government exemplifies and explains the continuities of industrial practice of pre- and post-war Germany, the major difference being that the post-war ideology of the social market economy has forced the phenomenon underground, so that it re-emerges in a largely informal guise. But the tradition is a very strong defining characteristic of German capitalism and has easily withstood the challenges of Allied attempts at decartelization and of neo-liberal economics. The concept and institution of industrial self-government embraces and unites the politics of the dominant state role and the economics of industrial production in an indissoluble whole, on the basis of self-regulated organization. But as the present study has amply demonstrated, the autonomy implied in the concept of self-organization is far from complete and in particular instances can be negligible. Major state involvement as a quantitative factor is irrefutable; it is the qualitative aspect which is of greater interest here.

The Central, Mediating Role of the Chambers

The business chambers, especially Chambers of Commerce, were the first formal institution through which private industrial affairs were self-regulated, whilst simultaneously becoming semi-state organs.[1] Their special organic relationship with the authorities 'found its clearest expression in the right of direct access to the executive branches of government'.[2] These public law bodies entertained intimate relations with other private groups:

> The so-called free or private associations were also bound to the system of institutionalized co-operation with state bodies through a variety of personal and institutional ties. In 1879 almost one third of the member associations of the Central Association of German Industrialists ... were Chambers of Commerce and in 1911 at least 31 of 53 committee members of the Association of German Chambers of Commerce were at the same time members of the Hansa League for Trade, Commerce and Industry, that is of a 'free' association. Correspondingly, over half of the Chambers of Commerce were corporate members of the Hansa League.[3]

The frequent and intense contact between these organizations contributed towards an 'erosion of the distinction between private and public sectors',[4] and later to that 'interpenetration of state and business interests which would characterize the German political economy during and after World War One'.[5]

What precisely were the attractions of such close association? It was certainly not a case of one-way traffic, with capitalist plutocrats moving in on the state machine, simply in order to make it more malleable to their own demands and interests: 'they were more interested by the security of institutionalized co-operation with the state and national administrations.'[6] The officials of nominally voluntary associations were to demonstrate a responsiveness 'to the ideal of the state bureaucrat and to the idea of the paternalist social state',[7] and soon 'the process of incorporation of free associations into the welfare state bureaucracy of Prussia-Germany was well-advanced'.[8]

Indeed, very often, 'the traffic' (i.e. the flow of influence) proceeded in the opposite direction, such that 'state intervention was mediated more and more by major groups in society'.[9] One consequence of this process was particularly noted by Maier: 'The Germans brought a model of civil service leadership into the factory: the state hierarchy legitimized the rankings of private enterprise.'[10] Private organizations generally acquired a public orientation, aspiring to acquire quasi-public powers. That this was partly a reflection of a universal bureaucratic power phenomenon was demonstrated by the fact that at the turn of the century the Farmers' League employed more officials than the Imperial Ministry of the Interior! Whatever the implications of widespread bureaucracy, by the outbreak of the First World War there already existed a large measure of self-government and interpenetration of state and business (both being defining features of organized capitalism and of corporatism). Events in Germany's premier industry at that time, the coal industry, particularly its cartelization, exercised a reciprocal influence on these developments. A foundation and pattern for inter-industry and, above all, for government–industry relations had been established which was to receive much modification in the course of time, but no fundamental alteration.

Organized Labour, Conciliation, and Corporatism

By the outbreak of the First World War, basically three kinds of self-governing bodies existed: the semi-public chambers, the industrial associations, and the cartels; in the latter two a large degree of genuine autonomy existed. But as has been demonstrated in Part I of this book within the mightiest cartel, the RWCS, the state interest had been increasing rapidly during the first decade of the twentieth century. During the First World War itself:

All in all, the self-governing bodies (e.g. especially the associations and the cartels) acquired the character of semi-public institutions. Towards the end of the war they were no longer the autonomous partners of government but agents controlled by it.[11]

After the war government influence was to recede gradually, but not to disappear; on the contrary, a new partnership emerged in which government became by no means the junior partner.

The two post-war eras were to evidence considerable similarities, not least in the relations between capital and labour. After the First World War the employers conceded the Central Working Community, legally binding collective bargaining, and the 8-hour working day, co-operated in establishing a commonweal economy solution for the coal industry, and acceded to the factory legislation of 1920 and 1922 creating works councils. However, the great inflation did not undermine or destroy the power of leading industrialists and their organizations; on the contrary, it furthered consolidation and amalgamation, finding its most concrete expression in the development of mighty trusts which in turn came to control the pervasive cartels. With the restoration of employer pre-eminence labour went into retreat. This was already foreshadowed by the SPD going into opposition, remaining out of power until 1928. The 8-hour day was lost, the Central Working Community and Imperial Economic Council degenerated into a talking shop, and labour influence within the new organs of the coal industry remained nominal.[12] By the end of the 1920s and the early 1930s it could be said that 'state steering is launched in that situation when private self-regulation breaks down due to the intensity of conflict of the participating social interests'.[13]

After another lost war German employers experienced a similar loss of prestige and authority. Again they resorted to what was intended as a brief alliance with organized labour against the common enemy, this time the Allied occupying power (previously, radical workerist socialism). The main concession, under Allied aegis being the granting of parity co-determination in what had become Germany's premier industry: iron and steel. Yet again, with the reconsolidation of private enterprise, as a result of the currency reform of 1948 and the election of a bourgeois coalition in 1949 (with the Allies also indicating a desire to relinquish certain *montan*-controls), the employers came off the defensive. Encouraged by indications that the Allies wished to abandon these controls the employers took up the offensive, wishing to withdraw, seemingly with government support, from parity co-determination. But on this occasion organized labour was to display a determination and unity of purpose, based on well-

CONCLUSION 229

organized, non-denominational and non-partisan *Einheitsgewerkschaften*, contrasting sharply with the divisions and debility of the 1920s. Parity co-determination was successfully defended, auguring well for a constructive interaction between capital and labour based on the necessary equality of standing. It was a positive augury because the *montan*-sector had, historically, displayed the worst industrial relations, thanks to the excessively authoritarian attitudes of employers.

The resolution displayed by the two unions, IG Metall and IG Bergbau and, of course, the DGB, in representing their respective work-forces, was not evidence of a fundamental sense of aggression or intransigence on behalf of organized labour, it merely reflected a dedication to avoid the mistakes and weaknesses of the past. This book has indicated that such a hostile stance would have been unlikely in view of labour's thoroughgoing commitment to the conciliatory ideology of economic democracy. The intellectual origins of the ideology lay with the reformist social democracy associated with the influence at the turn of the century of Eduard Bernstein which stressed co-operation as its principal feature. Just how the practice of the German labour movement was to demonstrate an active commitment to co-operation with both capital and state had been displayed at a most unlikely moment: during the period of so-called revolutionary ferment in the years immediately following the First World War. After rejecting nationalization the Socialization Commission[14] issued an interim report on 15 February 1919 claiming that:

> mutual understanding of arguments and motives finds expression as much in the fact that the groups eagerly participate in improving proposals which, on the whole, they do not accept as in both sides being prepared to support as a motion, in the event of their own draft being rejected, that of the remainder of the commission.[15]

A greater contrast with the approach of France's syndicalist-disposed unions could not be imagined. More importantly, if German labour could be so accommodating in such an environment, it is very difficult to envisage circumstances where it would not take such an approach.[16]

More than fifty-five years later, the leading labour movement theorist of economic democracy, Fritz Vilmar, was attempting to redefine it on the basis of the same co-operation of social groups, with the same corporatist assumptions. This is evident in his criticism of the deficiencies of the Concerted Action and of other instruments and boards created by the Stability Law of 1967 which possessed no institutional links and therefore also no opportunities to influence and vote on relevant issues.[17] Vilmar was also particularly critical of the nature and workings of the Stability Act itself, suggesting improvements to be expected from a national budgetary system. The DGB recommended that the preparation and development of the national budget should take place in committees in which the trade unions participate, along with other economic organizations. Participation in formulation of the national budget would represent the ultimate in incorporation and Vilmar agreed that procedures should engage all those affected, specifically: the workers, consumers, and representatives of the national interest (meaning the *Gemeinwohl*), political representatives or civil servants.[18] The intimate involvement of all interested parties in the general management of industry was considered particularly relevant in socialized industries, and this embraced the workforce and their trade union representatives, public administration, and consumer representatives. These proposals bear more than a passing resemblance to those of Dr Harald Koch (SPD), a former Economics Minister of the *Land* of Hesse, who in 1946 had developed an extremely detailed plan of an organization to run the enterprises taken into common ownership under the provisions of Article 41 of the (new) Hessian Constitution.

These social communities would each have been answerable to their own administrative council, composed equally of three interests: trade union officials, representing the workers; local government officers, representing consumers; and a *Land* Association of Social Communities (*Landesgemeinschaft*). The *Landesgemeinschaft*, with its twenty-eight members and expert committees, was not unlike the earlier Imperial Coal Council, but much more direct political control was envisaged, with the *Land* Minister-President as the chairman and the *Land* Economics and Transport Minister as vice-chairman.

A similar plan for the coal industry had also been evolved in 1946 by Dr Georg Berger of the mining union. A main, administrative mechanism of the projected social collieries would be a board of trustees, a *Treuhänderkollegium*, which would exercise trusteeship of the shares and within its ranks contain a sufficient number of representatives of the wider public interest. The voting rights of the voting shares would be exercised by the mining union, other associations, and by the works councils; also it was envisaged that the work-forces themselves would be able to acquire mining shares directly. This was certainly not nationalization and evinces further evidence of German labour's corporatist desire to engage all relevant interests.

The Real, Existing Social Market Economy

Yet Vilmar's systematic and extensive proposals would have been meaningless without the foundations of a re-established, robust corporatist structure upon which they could be grafted. Abelshauser succinctly describes the nature of the Federal German economic order, as it developed in the 1950s and 1960s:

> Even under Erhard, behind the façade of a market economic system the traditional method of sector-related self-steering by the industrial associations had developed further. This specifically German form of economic planning and steering was organized at association level, whereby the guidelines for development were formulated by the 'house bank' of the relevant sector.[19]

However, in a number of respects the coal industry represented a special case, not least because of the dramatic prospect of its sudden demise in the mid-1960s which extracted almost unprecedented financial support from the state. This support represented a paradigmatic case of the general principle that:

> If fundamental and wide-ranging decisions were to be taken then this model could easily be extended into a 'concerted action' of those immediately concerned, under the leadership of the ministerial bureaucracy.[20]

This is not to suggest that the state only becomes active in cases of crisis; investigation of the coal industry establishes that the active but normally informal nature of state involvement is often forced to become explicit and high profile. The role of the West German state in industrial affairs has been substantial and in some sectors, like energy, has been decidedly growing. It is the ideology of the social market economy which decrees that such activity is low profile, of an informal rather than of a formal nature. But the power of the German state tradition, not least as an honest broker between conflicting interests, has proved to be far stronger than the impact of neo-liberal theories, however rational and sophisticated their development and presentation.

Public policy-makers appear reluctant to admit the often decisive nature of their involvement and specific interventions, as is particularly evident in the energy sector where the conflict and contradiction between conventional market doctrines and the compulsions of the state tradition are at their sharpest. The official stance is unambiguous: 'energy policy is only designed to supplement the market mechanism and ... not a general regulation of all energy processes'.[21] Moreover, 'Altogether state influence in the energy field ... is ... not so great as to disturb or prevent the working of the free market economy or the basic structure of the energy industry orientated to competition and profit'.²²

But it is a nice irony that one of officialdom's examples used to demonstrate market pre-eminence actually contradicts the argument. 'On Federal Government level only one new institution has been created in the energy field, the Crude Oil Stockpiling Corporation.'²³ Apparently:

> the Stockpiling Corporation is characteristic of the energy sector in that it is an attempt to ensure that the free market forces can operate as far as possible in this sector despite the necessity for state control and state rights of intervention.²⁴

It was 'a body incorporated under public law subject to legal supervision by the Federal Ministry of Economics'.²⁵ The mostly private members possess 'a limited right of self-administration in the corporation . . . and . . . they can exercise an influence on the decisions as to how the corporation will fulfil its stockpiling obligation'.²⁶ But

> the Advisory Council representatives of the Federal Government have one third of the votes; . . . they cannot be outvoted by representatives of industry on major issues. Fundamental decisions, such as alterations to the statutes or regulations on membership contributions need the approval of the Economics Ministry.²⁷

Quite apart from the specific reference to self-government, it would seem that the foregoing information depicts the fundamental and decisive power of the state, the latter representing the ultimate last resort, and more.²⁸

What that other ultimate arbiter, the market, had come to mean for the coal industry was clearly expressed on behalf of the employers by Karl-Heinz Bund, managing director of the Ruhrkohle AG. He advocated a consensual approach to remedies for the energy sector whereby an unadulterated market was not the highest law but something to be supplemented by the participants. The corporatist approach becomes even more explicit with a 'consensus of all the partners directly participating in the development of the energy market, in other words the larger producer and consumer groups'. This would represent 'a market of "sensible opinions and intentions", alternatively "a market at the highest level" where prices operate as far as possible as a regulator'.²⁹ Belated official attempts to come to terms with the corporatist reality of Federal German industrial politics had begun in 1963 when an informal working party with representatives from the Federal Ministry of Economics and of the trade unions was formed. It recommended regular discussions, on the basis of national economic assessments, between those public authorities with economic responsibilities and 'the organizations of economic and social life'.[30] On 21 January 1966 a Social Dialogue began at the Federal Ministry of Economics involving members of the Council of Experts, various ministries and representatives from the trade unions, and the employers associations. This group of participants, whose membership was virtually identical with that of the later Concerted Action, continued its discussions in March of the same year until Erhard decided to discontinue them and their work was transferred to an unofficial group of experts. However, the major recession and the second coal crisis ushered in an entirely new phase in the Federal Republic's history, the corporatist nature of its political economy becoming more explicit, without necessarily becoming much more formalized. West Germany's corporatist, self-regulatory system was put to the test and was further consolidated, not rent asunder. Questions of the market and their ramifications were managed in a totally pragmatic way, particularly in the energy sector.

The Validity of Ronge's Three Alternatives

The first of Ronge's conditions for the successful materialization of non-state politics, seen in the form of private self-regulation, was represented by the capacity of individual companies to renounce a large degree of autonomy in the cause of solidaristic self-organization. Such behaviour entailed giving greater priority to certain considerations other than to an automatic, spontaneous response to market demands in the pursuit of profit. The second condition relates to an output by private parties which corresponds to a public purpose. The third condition entailed either preempting a state intervention or relieving the state of an additional burden of responsibility. Yet each of the three conditions can also represent a component, either separately or in combination, of a neo-corporatist analysis. They do not suffice to distinguish private self-regulation from its apparent opposite, state intervention, with corporatism as an intermediate mode of interaction.[31] Czada and Dittrich are forced to conclude

that the border between state and private forms of regulation runs neither in a straight line nor remains fixed at a certain level; its course varies much more according to the economic sectors and fields of conflict within the capitalist economy. The models of politicization posited by Ronge—state intervention, corporatism, private self-regulation—would accordingly be only partially 'interchangeable' and not possibilities of steering which can be employed at will, but ones tailored much more to specific areas and to particular situations or problems.[32]

Yet it is necessary to go beyond that common-sense finding, not least because 'a "binding policy-formulation" by private agents meets restrictions and limits at that point where far-reaching decisions are to be made'.[33] But, more importantly, Ronge's argument is based on two misconceptions. First, on a misunderstanding of the nature of the German tradition of self-government. Industrial self-government, as interpreted in this book, incorporates and transcends a self-regulation or self-organization which entails a minimal or non-existent state role: these represent minor variations on a major theme. The major theme is that of a self-government which has possessed a state dimension virtually from its inception. Self-government/regulation only represents an alternative to the state when the state is engaged in nationalization or detailed state regulation; experience has shown that private industrial self-government is compatible and consistent with a varying and often very substantial degree of state involvement.

Ronge's second misconception seems to concern an assumption of unambiguous interest. The case of the German coal industry has presented formidable ambiguities in the definition of interests particularly in the 1970s. What takes precedence in the Ruhrkohle AG's decision-taking in particular instances? Is it coal, steel, or electricity, not to mention chemicals? When is labour not labour? By being fully integrated into company management structures through the institution of parity co-determination labour has, in a sense, become an employer, although the union resolutely refuses to consider this implication. Given the integration of IG Bergbau into the politics of the SPD at all levels it is difficult to claim total separation from state activity, especially when the SPD is in power at Federal level.

Just as the theories of organized capitalism and neo-corporatism have accepted the blurring of distinctions between capital and the state, it is now necessary to consider the blurring between capital and labour and possibly even between labour and the state. Although it can be argued that the coal industry represents something of an extreme case, the study of its travails has demonstrated a huge degree of flexibility in the functioning of the West German political economy. The power of tradition has not represented a dead weight but has been a constant source of strength by ensuring continuity in industrial policy-making despite the superficial overlay of social market doctrines.

The present examination of the evolution of the (West) German coal-mining industry has led to the conclusion that it is the phenomenon of industrial self-government which has been primarily responsible for giving coherence, consistency, and continuity to industrial policy-making in the Germany of the pre- and post-war years. The coal employers may have demonstrated more cohesion than employers in other industries and the industry's special relationship with the state was probably not replicated elsewhere, but other factors ensured that coal did not develop in splendid isolation. It was integrated especially through its intrinsic importance as a supplier of an essential product; through the prestige its executives enjoyed in employers' organizations; and, most importantly, through the multiplicity of vital, intimate links of a technological and organizational kind with other industries, primarily iron and steel, chemicals, and the remainder of the energy sector. The influence and impact of the coal industry on other industries and upon industrial development generally, although not quantifiable, was very considerable. For more than one hundred years of German industrialization coal has been of absolutely central importance.

The institution of industrial self-government is characterized by a high degree of co-ordination among the employers within a particular industry and often between those of different industries (i.e. concertation at the expense of divisions within capital); by a close working relationship, i.e. co-operation, with the state; and, in the post-war era, by the integration of labour, primarily through industrial democracy, into the relevant decision-making processes. The term self-government, properly understood, does not exclude state participation. Conceptually and in its institutional manifestation self-government normally entails a state dimension, often of a substantial kind, not as a matter of inherent definitional logic but as a contingent historical fact.

The year 1945 may have represented a *Null-Punkt* in German political history; from an industrial viewpoint it certainly did not. In particular, there was no change in ownership relations, merely a temporary suspension in the *montan*-sector. Despite the more conclusive nature of the defeat, there was less organizational innovation after the Second than after the First World War. The biggest changes came in increased worker participation in management through the two Co-determination Acts of 1951 and 1952. Although dependent upon previous Allied initiatives, these Acts none the less built upon 1920s' precedents, with parity co-determination being introduced in steel which was to go into relative decline and in coal which was soon to go into absolute decline.

Nevertheless, increased worker participation, particularly in the *montan*-sector, was of great symbolic importance and its practical consolidation, assisted by the rise and establishment of an affluent society, exercised a stabilizing effect both on industrial relations and upon associated relations in the wider society. Devotees of the social market economy would argue that it was the successful application of its doctrines that was primarily the cause of the economic growth and social reconciliation. But it cannot be proven that the application of this ideology was a necessary, let along a sufficient, condition of German economic and political resurgence. Many other factors were involved which may have been more important—Germany had certainly experienced high economic growth before, under totally different socio-political regimes.

During the 1950s and 1960s the practice of co-determination appeared to reveal a compatibility with customary company management procedures in both the private and public sectors. In coal-mining, co-determination proved itself at pit-, at company-, and, with the foundation of the Ruhrkohle AG, virtually at industry-wide level. But, as the present study has indicated, there was more to labour integration than co-determination. Equally important was the creation of non-partisan industrial unions and, in the case of the mining union, almost total integration into the prevailing political system (partly associated with the inexorable post-war rise of the SPD). In addition, the West German trade union movement owned the holding company which had the most extensive holdings of any such company in the Federal Republic,[34] and exerted influence over VEBA, the huge energy conglomerate which had recently been a semi-state concern,[35] through the supervisory board deputy-chairman Adolf Schmidt, the head of IG Bergbau and Energie. In these circumstances, the Federal Republic clearly possessed a very special kind of market economy. Indeed, organized labour's manifold involvement could hardly have been more extensive.[36] But not only the extent but also the depth of that commitment has been illustrated by the case of the coal industry. The events associated with the second coal crisis, embedded in, and causally related to, the wider national political and economic crisis, have demonstrated how dependent the existing system is upon the active and constructive integration of organized labour; system survival depends on a positive labour contribution. This system will be subject to further testing with the integration of the two Germanies.

These findings can only be adequately understood in their historical context. In particular, the experience of war and its aftermath has materially influenced institutional and organizational arrangements not only politically but also in industrial and trade union affairs. It is highly unlikely that parity co-determination would have been established in steel and extended to coal, without the war and the ensuing Allied occupation. The very labour demand for parity co-determination, itself an essential component of full co-determination, cannot be fully appreciated without knowledge of the labour movement's previous experiences, particularly in the 1920s.

Clear recognition also has to be granted to the substantial, intimate, and possibly necessary role of the state in industrial affairs. The state plays a major role in maintaining system stability and growth by actively supporting and, where necessary, initiating industrial innovation, to maintain German economic pre-eminence in the material interest of all citizens, not just of capital. The role of the state in industrial affairs is not unlike that of the private banks, but the government is also the banker of the last resort, both in a literal and metaphoric sense of the term.[37]

The common failure to understand the contemporary role of the state in modern German capitalism induces certain bourgeois analysts to ignore or deny the continuities of industrial policy-making in the Federal Republic which are consistent with a tradition of *Selbstverwaltung*. Other observers fail to appreciate the power of traditional continuity in the policy-making and appear to take self-regulation more or less at face value, again ignoring or underestimating the true nature of the state contribution.[38] In contrast, it has become a central tenet of the present work that the evolution of West German politics as such and the evolution of many other major social institutions in Germany may perhaps best be described within the terms of the philosophy and organization of *Selbstverwaltung* (self-government).[39] Certainly the different modes of organization within the German political economy can be investigated and evaluated as variations on this central theme. The German coal industry represents the apotheosis of a multifaceted industrial self-government. The Ruhrkohle AG in particular embodies a modern redefinition of German industrial *Selbstverwaltung* in the contemporary context, drawing on a tradition more powerful than that of conventional market doctrines.

It may, finally, be argued that forces are at work in contemporary industrial societies that undermine existing forms of organized capitalism.[40] For German organized capitalism to collapse, at least two conditions would have to be met: a breakdown in consensus politics, and the disintegration of industrial *Selbstverwaltung* (industrial self-government/ organization/regulation). If these twin pillars of German corporatism collapsed, then an unorganized, if not disorganized, capitalism would almost certainly obtain. However, modern German capitalism is most emphatically neither unorganized nor disorganized, however much the forms and structures of organization change as circumstances change. The evidence is overwhelming that the ideological consensus is being sustained,[41] and also that German business and industrial organizations are particularly active in the merging of the two Germanies. It is not so much individual companies leading the way in establishing a market economy in former East Germany but the state, banks, and business chambers. The East is being appropriately reorganized to facilitate the necessary integration with the West German economy. An unorganized or disorganized German capitalism is virtually inconceivable: it would require a revolution in values and institutions which does not seem at all imminent. The German industrial system has been highly successful according to virtually any usual criteria, a complete transformation or deformation in the short to medium term which would jeopardize this appears, at the very least, improbable.

Notes

1. For an excellent account of their purpose and structure, see F. Glum, *Selbstverwaltung der Wirtschaft* (Berlin: Hermann Sack Verlag, 1925), 162–5.
2. W. Abelshauser, 'The First Post-Liberal Nation: Stages in the Development of Modern Corporatism in Germany', *European History Quarterly*, 14 (1984), 291.
3. Ibid. 291–2.
4. C. S. Maier, *Re-casting Bourgeois Europe* (Princeton, NJ: Princeton University Press, 1975), 11.
5. Ibid. 35.
6. Abelshauser, 'The First Post-Liberal Nation', 292.
7. Ibid.
8. Ibid.
9. Ibid. 290.
10. Maier, *Re-casting Bourgeois Europe*, 88.
11. R. Czada and W. Dittrich, 'Politisierungsmuster zwischen Staatsintervention und gesellschaftlicher Selbstverwaltung', in V. Ronge (ed.), *Am Staat vorbei* (Frankfurt: Campus Verlag, 1980), 210.
12. One advance retained by labour was state-sponsored arbitration in industrial relations. A further gain was the establishment of the Imperial Labour Office in 1927. But, in both cases, labour power rested upon state sanction, not on its own inherent strength.
13. Czada and Dittrich, 'Zwischen Staatsintervention und gesellschaftlicher Selbstverwaltung', 213.

14. The Commission contained, among others, the following members, Otto Hue, head of the Free Mining Union, top industrialist Walther Rathenau, bourgeois economists Alfred Weber and Josef Schumpeter, and socialist theoreticians Karl Kautsky and Rudolf Hilferding.
15. *Bericht der Sozialisierungskommission über die Frage der Sozialisierung des Kohlenbergbaus* (Berlin: Verlag Hans Robert Engelmann, 1920), 33. (But, nevertheless, both majority and minority reports were eventually issued.)
16. Even at the very beginning of the Third Reich, the official labour hierarchy went to considerable lengths to adapt and adjust promptly to the demands of the new rulers.
17. See F. Vilmar, 'Volkswirtschaftliche Rahmenplanung und Investitionslenkung', in id., *Theorie der Wirtschaftsdemokratie*, Course 2 of a series of lectures for the West German Open University (Fernuniversität), Hagen, 1980, 43.
18. Ibid. 50.
19. W. Abelshauser, 'Von der Kohlenkrise zur Gründung der Ruhrkohle AG', in U. Borsdorf and H. Mommsen (eds.), *Glück auf, Kameraden! Die Bergarbeiter und ihre Organisationen in Deutschland* (Cologne: Bund-Verlag, 1979), 426–7.
20. Ibid. 427.
21. R. Haupt, H. Mirbach, and H. Siedentopf, *The Adjustment of Administration to the Energy Crisis*, International Symposium Report, Brussels (Bonn: Federal Ministry of Economics, 9–11 June 1982), 22.
22. Ibid. 42.
23. Ibid. 6.
24. Ibid. 6–7.
25. Ibid. 7.
26. Ibid.
27. Ibid.
28. In lignite production: 'major projects such as the opening of new pits cannot be put in hand without agreement with political bodies' (ibid. 24). For gas supply, also, see ibid.
29. K. Bund, 'Preisbildung für die Steinkohle—Prinzipien und Praxis', *Glückauf*, 112: 24 (1976), 1385.
30. W. Bonss, 'Gewerkschaftliches Handeln zwischen Korporatismus und Selbstverwaltung—Die Konzertierte Aktion und ihre Folgen', in Ronge, *Am Staat vorbei*, 134.
31. Ronge's view of corporatism contrasts, for example, with Schmitter's interpretation and with Lehmbruch's 'liberal corporatism'. Cf. E. Wyn Grant, *The Political Economy of Corporatism* (London: Macmillan, 1985).
32. Czada and Dittrich, 'Zwischen Staatsintervention und gesellschaftlicher Selbstverwaltung', 195.
33. Ibid. 216.
34. Not to mention also the tenth largest bank, the Bank für Gemeinwirtschaft (BfG) and the largest civil construction company in the free world, Neue

Heimat. (The BfG has since been sold to an insurance consortium and Neue Heimat liquidated.)
35. VEBA was privatized by the Kohl government during the 1980s. This innovation did not represent a fundamental change in either energy or coal policy.
36. Although it did not introduce parity co-determination into its own enterprises until the early 1970s.
37. The role of labour representatives (particularly those of the trade unions) on company boards in declining industries is also not unlike that of those of the banks in prospering industries: the unions appear to play a compensatory role here.
38. It seems to be a frequent error to assume that if in a particular instance of industrial reorganization the state plays a subordinate role that role is necessarily minor or even peripheral; such is by no means always the case, e.g. the rescue of AEG in the early 1980s.
39. A *Selbstverwaltung* which also reflects the fact that: 'In contrast to the pluralistic, competitive model of Anglo-Saxon stamp, the *Proporz* model of political conflict regulation creates favourable preconditions, ideologically and organizationally, for a political self-government of private associations' (Czada and Dittrich, 'Zwischen Staatsintervention und gesellschaftlicher Selbstverwaltung', 203).
40. For a representative view, see S. Lash and J. Urry, *The End of Organized Capitalism* (Cambridge: Polity Press, 1987).
41. The incorporation of ecological politics, e.g. the Green Party, into the established politico-economic system is a further manifestation of its inexorable power.

Bibliography

ABELSHAUSER, W., 'Von der Kohlenkrise zur Gründung der Ruhrkohle AG', in U. Borsdorf and H. Mommsen (eds.), *Glück auf, Kameraden! Die Bergarbeiter und ihre Organisationen in Deutschland* (Cologne: Bund-Verlag, 1979).
—— 'Korea, die Ruhr und Erhards Markwirtschaft: Die Energiekrise von 1950/51', *Rheinische Vierteljahresblätter*, 45 (1981).
—— 'Ansätze korporativer Marktwirtschaft in der Koreakrise der frühen fünfziger Jahre', *Vfz*, 30 (1982), Heft 4.
—— *Der Ruhrbergbau seit 1945* (Munich: Verlag C. H. Beck, 1984).
—— The First Post-Liberal Nation: Stages in the Development of Modern Corporatism in Germany', *European History Quarterly*, 14 (1984).
ADEMSEN, H. R., *Investitionshilfe für die Ruhr* (Wuppertal: Peter Hammer Verlag, 1981).
ADLER, F., *Forschung und Entwicklung im Steinkohlenbergbau* (Expert investigation report), sponsored by Ministerium für Wirtschaft, Mittelstand und Verkehr, Düsseldorf, Feb. 1966.
—— 'Steinkohle', in G. Bischoff and W. Gocht (eds.), *Das Energiehandbuch* (Braunschweig: Vieweg & Sohn Verlagsgesellschaft, 3rd edn., 1979).
Aktionsgemeinschaft Deutsche Steinkohlenreviere GmbH, *Annual Report* (*Geschäftsbericht*) for 1979 (Düsseldorf: Verlag Glückauf, 1980).
ALBERT, W., 'Die Wiedereingliederung der westdeutschen Kohlenwirtschaft in die Marktwirtschaft', Ph.D. thesis, Rheinische Friedrich-Wilhelms-Universität, Bonn, 1961.
ALMANASREH, Z., 'Institutional Forms of Worker Participation in the Federal German Republic', in D. Heathfield (ed.), *The Economics of Co-determination* (London: Macmillan, 1977).
ANDERHEGGEN, E., *et al.*, *Mineralische Rohstoffwirtschaft* (Bonn-Bad Godesberg: Neue Gesellschaft, 1971).
ARENDT, W., 'Mitbestimmung ist für die SPD paritätische Mitbestimmung', *Mitbestimmungsgespräch*, 11 (1970).
Ausschuß zur Untersuchung der Erzeugungs- und Absatzbedingungen der deutschen Wirtschaft (III Unterausschuß), *Die deutsche Kohlenwirtschaft* (Berlin: Verlag E. S. Mittler u. Sohn, 1929).
BACKES, W., 'Die Entwicklung der Arbeitsproduktivität im deutschen Steinkohlenbergbau 1956–1970', Ph.D. thesis, Rheinische Friedrich-Wilhelms-Universität, Bonn, 1974.
BAHL, V., *Staatliche Politik am Beispiel der Kohle* (Frankfurt: Campus Verlag, 1977).
BALS, H., and SCHNEIDER, H. K., 'Energiepolitik statt Kohlenpolitik', in *Hamburger Jahrbuch für Wirtschafts- und Gesellschaftspolitik*, 13 (Tübingen, 1968).

BAUER, A., 'Die Bedeutung des Steinkohlenbergbaus für die konjunkturelle Entwicklung der westdeutschen Wirtschaft', Ph.D. thesis, Hochschule für Wirtschafts- und Sozialwissenschaften, Nuremberg, 1957.
BAUMGARTEN, H., *Der deutsche Liberalismus* (Frankfurt: Verlag Ullstein, 1974).
BEER, W., *Mitbestimmung* (Bochum: IG Bergbau und Energie, 1975).
BENNECKE, P., *Die Subventionspolitik der Hohen Behörde der europäischen Gemeinschaft für Kohle und Stahl und ihre Auswirkungen auf den Kohlenbergbau dieser Gemeinschaft* (Cologne: Westdeutscher Verlag, 1965).
BERDING, H., et al., *Organisierter Kapitalismus* (Göttingen: Vandenhoeck & Ruprecht, 1974).
Bericht der Sozialisierungskommission über die Frage der Sozialisierung des Kohlenbergbaus (Berlin: Verlag Hans Robert Engelmann, 1920).
BIEDENKOPF, K. H., *Thesen zur Energiepolitik* (Heidelberg: Verlagsgesellschaft 'Recht und Wirtschaft', 1967).
BLACKABY, F. (ed.), *De-industrialisation* (London: Heinemann, 1978).
BLUME, O., DUVERNELL, H., and POTTHOFF, E., *Zwischenbilanz der Mitbestimmung* (Tübingen: J. C. B. Mohr (Paul Siebeck), 1962).
Board of Management, Ruhrkohle AG, *Neuordnung des Bergbaus* (Essen: Nov. 1969).
BOLDT, G., *Staat und Bergbau* (Munich: C. H. Beck' sche Verlagsbuchhandlung, 1950).
BORCHARDT, K., 'The Federal Republic of Germany', in G. Stolper, *The German Economy: 1870 to the Present* (London: Weidenfeld & Nicolson, 1967).
——*Die industrielle Revolution in Deutschland* (Munich: R. Piper & Co. Verlag, 1972).
BORSDORF, U., 'Wirtschaftdemokratie und Mitbestimmung: Historische Stufen der Annäherung an den Kapitalismus', *WSI-Mitteilungen*, 3 (1986).
——and MOMMSEN, H. (eds.), *Glück auf, Kameraden! Die Bergarbeiter und ihre Organisationen in Deutschland* (Cologne: Bund-Verlag, 1979).
BOWDEN, P., *Regional Development in Action* (North-Rhine Westphalia: North West England; London: Anglo-German Foundation, 1979).
BRAUBACH, B., *Coal Subsidy Policy in the Federal Republic of Germany* (Bonn: Federal Ministry of Economics, May 1986).
BRECHT, A., and GLASER, C., *The Art and Technique of Administration in German Ministries* (Westport, Conn.: Greenwood Press, 1971; orig. Cambridge, Mass.: Harvard University Press, 1940).
BREDER, H., *Subventionen im Steinkohlenbergbau* (Berlin: Duncker & Humblot, 1958).
BRIEFS, G. (ed.), *Mitbestimmung?* (Stuttgart: Seewald Verlag, 1967).
BRINCKMAN-HERZ, D., *Die Unternehmensmitbestimmung in der Bundesrepublik Deutschland* (Cologne: Kiepenhauer & Witsch, 1975).
BUCHHOLZ, K., et al., *Mitbestimmung und Interessenvertretung* (Cologne: Bund-Verlag, 1981).

BUDDEE, K., 'Determinanten und Entwicklungstendenzen des Absatzes der Ruhrkohle AG und die Problematik einer absatzorientierten Produktion', Ph.D. thesis, Wirtschafts- und Sozialwissenschaftliche Fakultät der Universität zu Köln, Cologne, 1974.

BUND, K., 'Chancen, Risiken und Aufgaben der Kohle in einer veränderten Energiewirtschaft', *Glückauf*, 112: 1 (1976).

——'Preisbildung für die Steinkohle—Prinzipien und Praxis', *Glückauf*, 112: 24 (1976).

——'Steinkohlenbergbau als unternehmerische Aufgabe', *Glückauf*, 115: 12 (1979).

——'Die neue Rolle der Steinkohle', *Forum Energie*, 1 (Oct.–Nov. 1979).

BURCKHARDT, H. (ed.), *Deutscher Steinkohlenbergbau im Spannungsfeld zwischen Politik und Wirtschaft* (Cologne: Deutscher Industrieverlag, 1968).

——'Rohstoffversorgung und Rohstoffpolitik', *Glückauf*, 110: 1 (1974).

——*25 Jahre Kohlepolitik* (Baden-Baden: Nomos Verlagsgesellschaft, 1981).

BURGBACHER, F., *Die Energiesituation in der Bundesrepublik und die Zukunftsaussichten der Kohle* (Cologne: Westdeutscher Verlag, 1964).

——(ed.), *Ordnungsprobleme und Entwicklungstendenzen in der deutschen Energiewirtschaft* (Festschrift für Theodor Wessels) (Essen: Vulkan-Verlag, Dr W. Classen, 1967).

——MULLER, D., and WESSELS, T., *Die Energiewirtschaft im Gemeinsamen Markt* (Baden-Baden: Verlag August Lutzeyer, 1963).

BURTENSHAW, D., *Economic Geography of West Germany* (London: Macmillan, 1974).

BUSSIECK, J., 'Die betrieblichen Auswirkungen der Europäischen Gemeinschaft für Kohle und Stahl auf den westdeutschen Steinkohlenbergbau', Ph.D. thesis, Ludwig-Maximilians-Universität, Munich, 1958.

CALAMY, R., 'Organisierter Kapitalismus und Mitbestimmung', *Blätter für deutsche und internationale Politik*, 13: 11 (Nov. 1968).

CASSEL, D., GUTMANN, G., and THIEME, H. J. (eds.), *25 Jahre Marktwirtschaft in der Bundesrepublik* (Stuttgart: Gustav Fischer Verlag, 1972).

CAWSON, A., *et al.*, *Hostile Brothers* (Oxford: Clarendon Press, 1990).

CHANDLER, G., 'Energy, the International Compulsions', *Coal and Energy Quarterly*, 1 (Summer 1974).

CHILDS, D., and JOHNSON, J., *West Germany: Politics and Society* (London: Croom Helm, 1981).

CHRISTMAN, A., 'Allgemeine Probleme der Strukturpolitik', *Beihefte der Konjunkturpolitik: Strukturprobleme und ihre wirtschaftspolitische Bewältigung*, 16 (1969).

CLAPHAM, J. H., *The Economic Development of France and Germany, 1815–1914* (Cambridge: Cambridge University Press, 4th edn., 1968).

CLARK, J., *et al.*, *Trade Unions, National Politics and Economic Management* (London: Anglo-German Foundation, 1980).

CLOUGH, S. B., MOODIE, C., and MOODIE, T., *Economic History of Europe: Twentieth Century* (London: Macmillan, 1969).

Coal Employers' Association Annual Reports (Ruhrbergbau 1964–6; Steinkohle 1973/4, 1974/5, 1979/80, 1980/1, and 1988/9) (Essen: Verlag Glückauf).

Coalmining: Report of the Technical Advisory Committee (Reid Report), Cmd. 6610 (London: HMSO, 1945).

Co-determination Commission, *Mitbestimmung im Unternehmen* (Biedenkopf Report), Drucksache VI/334 (Bonn-Godesberg: Verlag Dr Hans Heger, 1970).

Commission for Economic and Social Change, *Wirtschaftlicher und sozialer Wandel in der Bundesrepublik* (Gutachten Report) (Göttingen: Verlag Otto Schwartz, 1977).

COMMONER, B., *The Poverty of Power* (London: Jonathan Cape, 1976).

CRABBE, D., and McBRIDE, R., *The World Energy Book* (London: Kogan Page, 1978).

DACH, G., and JAMME, H. P., 'Ein Jahr der Wende im europäische Steinkohlenbergbau', *Glückauf*, 111: 6 (1975).

——and —— 'Auswirkungen der wirtschaflichen Rezession auf den europäischen Steinkohlenbergbau', *Glückauf*, 112: 6 (1976).

DAHEIM, H., 'The Practice of Co-determination on the Management of German Enterprises', in W. Albeda (ed.), *Participation in Management: Industrial Democracy in Three West European Countries* (Rotterdam: Rotterdam University Press, 1973).

DAHRENDORF, R., *Class and Conflict in Industrial Society* (London: Routledge & Kegan Paul, 1957).

DEIST, H., *Die Neuordnung in der Montanwirtschaft und die Mitbestimmung in den Holding-Gesellschaften* (Bochum: Verlagsgesellschaft der IG Bergbau, 1954).

——and HESS, O., *Gutachten über die Kosten- und Ertragslage des westdeutschen Steinkohlenbergbaus* (Essen: Verwaltung für Wirtschaft des vereinigten Wirtschaftsgebietes, Mar. 1949).

DENTON, G., *Economic Planning and Policies in Britain, France and Germany* (London: George Allen & Unwin, 1968).

DEPPE, F., et al., *Kritik der Mitbestimmung* (Frankfurt: Suhrkamp Verlag, 1969).

DEUBNER, C., 'Change and Internationalization in Industry: Toward a Sectoral Interpretation of West German Politics', *International Organization*, 38 (1984).

DIEL, R., RADTKE, G., and STOSSEL, R., *Investment Requirements in the Energy Sector and their Financing* (Frankfurt: Dresdner Bank, 1980).

DOLINSKI, U., and ZIESING, J., *Sicherheits- Preis- und Umweltaspekte der Energieversorgung* (Berlin: Duncker & Humblot, 1976).

Dresdner Bank, *Development in Federal Germany's Energy Industry to 1980/85* (Frankfurt: Dresdner Bank, 1974).

DROSTE, M., 'Die Stellung des Ruhrbergbaus in Staat und Gesellschaft bis zum Jahre 1918', Ph.D. thesis, University of Göttingen, 1953.

DYSON, K. H. F., 'The Ambiguous Politics of Western Germany: Politicisation in a "State" Society', *European Journal of Political Research*, 7: 4 (Dec. 1979).

—— *The State Tradition in Western Europe* (Oxford: Martin Robertson, 1980).

—— 'The Politics of Economic Management in West Germany', in W. E. Paterson and G. Smith (eds.), *The West German Model: Perspectives on a Stable State* (London: Frank Cass, 1981).
—— 'Economic Policy', in G. Smith, W. E. Paterson, and P. H. Merkl (eds.), *Developments in West German Politics* (London: Macmillan, 1989).
—— and WILKS, S. (eds.), *Industrial Crisis: A Comparative Study of the State and Industry* (Oxford: Martin Robertson, 1983).
ECKSTEIN, D., *Die wirtschaftliche Betätigung der öffentlichen Hand im Bergbau und in der Elektrizitätswirtschaft der Bundesrepublik Deutschland* (Stuttgart: Gustav Fischer Verlag, 1966).
Economic History Review, 1–5, 8, 9, 11–13 (1927–43); 2nd ser.: 21–4 (1968–71), 26–8 (1973–5) (London: Economic History Society, 1927–).
EHLEN, K. J., *Die Filialgroßbanken* (Stuttgart: Gustav Fischer Verlag, 1960).
ELLWARDT, M., *Demokratischer Staat—Verbändestaat—Staat der Monopole?* (Marburg: Verlag Arbeiterbewegung und Gesellschaft, 1983).
ELSENHANS, H., 'Die weltwirtschaflichen Folgen der Energiekrise', *Sozialwissenschaftliche Information für Unterricht und Studium*, 9: 4 (Oct. 1980).
Energy Report of the Government of the Federal Republic of Germany, 24 Aug. 1986 (Bonn: Press Dept., Federal Chancellor's Office, 1990).
ENGEL, C., 'Industrielle Energiekosten und regionale Strukturpolitik', *Oel*, 2 (1967).
Erklärung der Bundesregierung zu den Kohlepolitischen Verhandlungen mit der EG-Kommission (Bonn: Press Dept., Federal Chancellor's Office, 1990).
ESCHENBURG, T., *Zur politischen Praxis in der Bundesrepublik* (Munich: R. Piper & Co. Verlag, 1964).
European Economic Community, *Die Anpassung der Arbeitnehmer und Umstellung der Gebiete* (Brussels: EEC Press & Information Office, 1967).
EVANS, D., *The Politics of Energy* (London: Macmillan, 1976).
—— *Western Energy Policy* (London: Macmillan, 1978).
Federal Coal Commissioner, *Untersuchung über die Entwicklung der Unternehmensgrössen im deutschen Steinkohlenbergbau* (Bonn: Federal Ministry of Economics, 1968).
Federal Ministry of Economics, *Entwicklung des deutschen Steinkohlenbergbaus* (Bonn: 1970).
—— *Gründung der Ruhrkohle AG (RAG) im Rahmen der Neuordnung des Steinkohlenbergbaus der BRD* (Bonn: 1970).
—— *Die Kohle und die Reviere haben eine Zukunft*, Foreword by Minister Schiller (Bonn: May 1970).
—— *Zeittafel der Ruhrkohle AG* (Bonn: 1973).
—— *Energieprogramm der Bundesregierung: Zweite Fortschreibung vom 14.12.1977* (Bonn: Dec. 1977). English trans., published by Federal Ministry of Economics, *Energy Policy Programme for the Federal Republic of Germany: Second Revision* (14 Dec. 1977).

Federal Ministry of Economics, *Entwicklung des Ruhrreviers* (Bonn: 1979).

——*Die Politik der Bundesrepublik Deutschland zur rationellen und sparsamen Energieverwendung und zur Substitution von Öl/Policy of the Federal Republic of Germany to Promote Rational and Economic Use of Energy and to Replace Oil* (Bonn: 1981).

Federal Press and Information Office, *Grundlinien und Eckwerte für die Fortschreibung des Energieprogramms, Bulletin*, 30 (1977).

——'Kontinuierliche Energiepolitik und verstärktes Sparen', *Bulletin*, 65 (1979). English edn., *Continuing Energy Policy and Enhanced Energy Conservation*.

——Chancellor Schmidt, 'Erklärung der Bundesregierung zur Energiepolitik nach dem Europäischen Rat und dem Weltwirtschaftsgipfel', *Bulletin*, 87 (1979).

——'Maßnahmen der Bundesregierung zur Energiesparung', *Bulletin*, 107 (1979).

——'Wirtschaftspolitische Folgerungen aus der Ölverknappung', *Bulletin*, 155 (1979).

——'Regierungserklärung des Bundeskanzlers vor dem Deutschen Bundestag', *Bulletin*, 124 (1980).

FLETCHER, W. J., and LENIHAN, J. (eds.), *Energy Resources and the Environment* (Glasgow: Blackie, 1975).

FORSTER, K., *Allgemeine Energiewirtschaft* (Berlin: Duncker & Humblot, 2nd edn., 1973).

FRIEDENSBURG, F., 'Rationalisierungsmaßnahmen und Rationalisierungserfolge im Steinkohlenbergbau der Bundesrepublik', *Vierteljahreshefte zur Wirtschaftsforschung*, 3 (1967).

Friedrich Ebert Stiftung, *Die Energiewirtschaft* (Hanover: Verlag für Literatur und Zeitgeschehen, 1960).

GABLER, J., 'Participation and the Working of the Price Mechanism in a Market Economy', in D. F. Heathfield (ed.), *The Economics of Co-determination* (London: Macmillan, 1977).

GAFGEN, G., *Grundlagen der Wirtschaftspolitik* (Cologne: Kiepenhauer & Witsch, 1966).

GEBHARDT, G., *Ruhrbergbau: Geschichte, Aufbau und Verflechtung seiner Gesellschaften und Organisationen* (Essen: Verlag Glückauf, 1957).

GEIBLER, A., and RIEGERT, B., *Energiepolitik für eine lebenswerte Zukunft* (Bonn: Verlag Neue Gesellschaft, 1988).

GILLINGHAM, J., *Industry and Politics in the Third Reich: Ruhr Coal, Hitler and Europe* (London: Methuen, 1989).

GLUM, F., *Selbstverwaltung der Wirtschaft* (Berlin: Hermann Sack Verlag, 1925).

GORDON, R. L., *The Evolution of Energy Policy in Western Europe: The Reluctant Retreat from Coal* (New York: Praeger, 1970).

GRATWOHL, M., *Energieversorgung* (Berlin: Walter de Gruyter, 1978).

GRIFFIN, A. R., *The British Coalmining Industry* (Buxton: Moorland Publishing Co., 1977).

GROHMANN, H., 'Veränderungen der Unternehmensstruktur im deutschen Steinkohlenbergbau in den Jahren 1850 bis 1969', *Glückauf*, 106: 8 (1970).
HAASE, W., *Formen regionaler Wirtschaftsförderung im Zusammenhang mit der Gründung der Ruhrkohle AG* (Cologne: Carl Heymanns Verlag KG, 1972).
Handelsblatt, *Zwang und Grenzen der Konzentration* (Düsseldorf: 1966).
HARTRICH, E., *The Fourth and Richest Reich* (London: Collier Macmillan, 1980).
HASELOFF, W., *Die politischen Parteien* (Frankfurt: Verlag Moritz Diesterweg, 1972).
HAUPT, R., MIRBACH, H., and SIEDENTOPF, H., *The Adjustment of Administration to the Energy Crisis*, International Symposium Report, Brussels (Bonn: Federal Ministry of Economics, 9–11 June 1982).
HAUSSMAN, F., *Der Neuaufbau der deutschen Kohlenwirtschaft im internationalen Rahmen* (Munich: C. H. Beck' sche Verlagsbuchhandlung, 1950).
—— *Die öffentliche Hand in der Wirtschaft* (Munich: Verlag C. H. Beck, 1954).
HAWNER, K. H., 'Fortschritte im deutschen Steinkohlenbergbau—Erfolgsbilanz und Aussichten in Bergtechnik und Kohlenveredelung', *Glückauf*, 110: 2 (1974).
HEFFTER, H., *Die deutsche Selbstverwaltung im 19. Jahrhundert* (Stuttgart: K. F. Koehler Verlag, 2nd edn., 1969).
HEINRICHSBAUER, A., *Der Ruhrbergbau in Vergangenheit, Gegenwart und Zukunft* (Essen: Verlag Glückauf GmbH, 1948).
HELMRICH, W., *Das Ruhrgebiet: Wirtschaft und Verflechtung* (Münster: Aschendorffsche Verlagsbuchhandlung, 2nd edn., 1949).
HEMPEL, G., *Die deutsche Montanindustrie* (Essen: Vulkan-Verlag Dr W. Classen, 1969).
HENDERSON, W. O., *The State and the Industrial Revolution in Prussia, 1740–1870* (Liverpool: Liverpool University Press, 1958).
—— *The Industrial Revolution on the Continent* (London: Frank Cass, 1967).
—— *The Industrialization of Europe: 1780–1914* (London: Thomas & Hudson, 1969).
—— *The Rise of German Industrial Power, 1834–1914* (London: Temple Smith, 1975).
HENNIG, H., *Entflechtung und Neuordnung der westdeutschen Montanindustrie unter besonderer Berücksichtigung der Verbundwirtschaft zwischen Kohle und Eisen* (Berne: Verlag A. Franke, 1952).
HERGT, S. (ed.), *Mitbestimmung* (Opladen: Heggen-Verlag, 1974).
HOFFMAN, K. H., ANDERS, H., and VOIGT, F. (eds.), *Energie: Leistungen, Prognosen, Alternativen* (Mannheim: Mannheimer Verlagsanstatt/Verlagsanstalt Courier, 1972).
HOLLER, K. H., 'Kohlenwirtschaftspolitik in der Bundesrepublik', Diss., Westfälische Wilhelms-Universität, Münster, 1960.
IG Bergbau, *Energiepolitik statt Kohlenkrise (1957–1960)* (Bochum: 1960).
IG Bergbau und Energie, *Jahrbuch 1978/79* (Bochum: Berg-Verlag, 1980).

IG Bergbau und Energie, *Überbrückungskonzept für den deutschen Steinkohlenbergbau* (Cologne: IGBE, Aug. 1987).

IMBUSCH, H., *Arbeitsverhältnis und Arbeiterorganisationen im Deutschen Bergbau* (Berlin: J. H. W. Dietz, 1980; orig. published by Verlag des Gewerkvereins christlicher Bergarbeiter Essen-Ruhr, 1908).

Imperial Ministry of the Interior, *Verhandlungen, kontradiktorische, über deutsche Kartelle*, i–iv (Berlin: Reichamt des Innern, 1905).

Industrieverband Bergbau (IG Bergbau und Energie), *Informationsblatt* (union journal) (26 Jan. 1948–12 Dec. 1950).

ISSING, O., *Investitionslenkung in der Marktwirtschaft?* (Göttingen: Vandenhoeck & Ruprecht, 1975).

JABLONOWSKI, H. W., 'Gesellschaftliche Kooperationsformen und politisches Instrumentarium zur Bewältigung der Strukturkrise im Steinkohlenbergbau und des energiewirtschaftlichen Strukturwandels in der Bundesrepublik bis Anfang der 70er Jahre', Diss., Dortmund University, 1977.

JACKSON, M. P., *The Price of Coal* (London: Croom Helm, 1974).

JAEGGI, U., *Macht und Herrschaft in der Bundesrepublik* (Frankfurt: Fischer Bucherei, 1971).

Jahrbuch für Bergbau, Energie, Mineralöl und Chemie, 1970 (Essen: Glückauf, 1970), 653.

Jahreswirtschaftsbericht 1981 der Bundesregierung (Bonn: Drucksache 51/81, 29 Jan. 1981).

JAKLI, Z., 'Staatliche Intervention und private Politik im Energiesektor', in V. RONGE (ed.), *Am Staat Vorbei* (Frankfurt: Campus Verlag, 1980).

JAMME, H. P., 'Das neue gemeinschaftliche System von Beihilfen für den Steinkohlenbergbau der Gemeinschaft', *Glückauf*, 107: 6 (1971).

JENSEN, W. G., *Energy in Europe, 1945–1980* (London: Foulis, 1967).

JEWKES, J., *Public and Private Enterprise* (London: Routledge & Kegan Paul, 1964).

JUDITH, R., et al., *Montanmitbestimmung: Dokumente ihrer Entstehung* (Cologne: Bund-Verlag, 1979).

——*Montanmitbestimmung: Geschichte, Idee, Wirklichkeit* (Cologne: Bund-Verlag, 1979).

KAACK, H., and ROCH, R. (eds.), *Handbuch des deutschen Parteiensystems* (Opladen: Leske & Buderich GmbH, 1980).

KEMP, T., *Industrialization in Nineteenth-Century Europe* (London: Longmans, 1969).

KERSTAN, F., 'Die finanziellen Maßnahmen der Mitgliedstaaten der Europäischen Gemeinschaft zugunsten des Steinkohlenbergbaus im Jahre 1970', *Glückauf*, 106: 21 (1970).

KEYSER, T. 'Die Bildung von "Einheitsgesellschaften" im Ruhrbergbau nach dem Gesetz Nr. 27', in *Jahrbuch des deutschen Bergbaus* (Essen: Verlag Glückauf, 1952).

——'Gedanken über Bergbaupolitik', *Volkswirt*, 37 (1954).

KITSCHELT, H., *Kernenergiepolitik* (Frankfurt: Campus Verlag, 1980).
KLONNE, A., *Die deutsche Arbeiterbewegung* (Düsseldorf: Eugen Diederichs Verlag, 1980).
KOST, H., 'Die Entwicklung des Bergbaus seit 1945', *Volkswirt*, 37 (1954).
—— 'Die Tätigkeit der Deutschen Kohlenbergbauleitung-Schlußbereicht, *Glückauf*, 90 (1954).
KOSTLER, R., 'Die Praxis der Weimarar Betriebsräte im Aufsichtsrat', *Die Mitbestimmung/Demokratie*, 8 and 9 (1986).
KRAUTER, U. 'Energiepolitik', in Graf O. Vitzthum (ed.), *Rechtliche und Okonomische Probleme der Energie- und Rohstoffversorgung der Bundesrepublik Deutschland* (Munich: Hochschule der Bundeswehr, 1979).
KRUSE, J., 'Energiewirtschaft der BRD im Wandel', *Wirtschaftskonjunktur*, 24: 1 (1972).
KUDA, R. F., 'Mitbestimmung und organisierter Kapitalismus', *Gewerkschaftliche Monatshefte*, 20: 2 (1969).
LAMBSDORFF, O. Graf, *Zielsetzung* (Düsseldorf: Econ Verlag, 1977).
LAMPERT, H., *Die Wirtschafts- und Sozialordnung der Bundesrepublik* (Munich: Gunter Olzog Verlag, 1976).
Landeszentrale für politische Bildung Nordrhein-Westfalen, *Demokratische Gesellschaft: Konsensus und Konflikt*, i and ii (Munich: Gunter Olzog Verlag, 1975).
Land Government, *Nordrhein-Westfalen Programm 1975* (Düsseldorf: Mar. 1970).
—— *Landesentwicklungsberichte 1974, 1976, & 1979*, 38, 39, and 42; *Landesentwicklungsplan VI 1979* (Düsseldorf: 1974, 1976, and 1979).
—— *Politik für das Ruhrgebiet: Das Aktionsprogramm* (Sept. 1979).
LANTZKE, U., 'Expanding World Use of Coal', *Foreign Affairs*, 58: 2 (Winter 1979/80).
LASH, S., and URRY, J., *The End of Organized Capitalism* (Cambridge: Polity Press, 1984).
LAUSCHKE, K., *Schwarze Fahnen an der Ruhr* (Marburg: Verlag Arbeiterbewegung und Gesellschaftswissenschaft, 1984).
LEHMANN, K. H., and HEINSIUS, T., *Aktienrecht und Mitbestimmung* (Niederkassel-Mondorf: Titz Verlag, 1979).
LEVY, H., *Industrial Germany: A Study of its Monopoly Organisations and their Control by the State* (London: Frank Cass, 1966; orig. published Cambridge: Cambridge University Press, 1935).
LIEBERMANN, S., *The Growth of the European Mixed Economies, 1945–1970* (London: J. Wiley & Sons, 1977).
LIESEGANG, H. C. F. (ed.), *Gewerkschaften in der Bundesrepublik Deutschland* (Berlin: Walter de Gruyter, 1975).
LUCAS, N. J. D., *Energy and the European Communities* (London: Europa Publications, 1977).

MABRO, R. (ed.), *World Energy: Issues and Policies* (Oxford: Oxford University Press, 1980).
MAIER, C. S., *Re-casting Bourgeois Europe: Stabilization in France, Germany and Italy in the Decade after World War I* (Princeton, NJ: Princeton University Press, 1975).
MARKOVITS, A. S., *The Politics of the West German Trade Unions* (Cambridge: Cambridge University Press, 1986).
MARSHALL, A., *Industry and Trade*, Books I and II (London: Macmillan, 1923).
MARTINY, M., and SCHNEIDER, H. J. (eds.), *Deutsche Energiepolitik seit 1945, Vorrang für de Kohle: Dokumente und Materialien zur Energiepolitik der Industriegewerkschaft Bergbau und Energie* (Cologne: Bund-Verlag, 1981).
MAYER, F., 'Die Neuordnung des Ruhrkohlenbergbaus', *Mitbestimmungsgespräch*, 2 (1969).
MAYNTZ, R., and SCHARPF, F. W., *Policy-Making in the German Federal Bureaucracy* (Barking: Elsevier, 1975).
MELLOR, R., *The Two Germanies: A Modern Geography* (London: Harper & Row, 1978).
MEYER-RENSCHHAUSEN, M., *Energiepolitik in der Bundesrepublik Deutschland von 1950 bis heute* (Cologne: Pahl-Rugenstein Verlag, 1977).
MIKAT, P., *Interim Report/Coal Commission* (Essen: Ruhrgas AG, on behalf of Federal Ministry of Economics, 1990).
MILWARD, A., and SAUL, S. B., *The Development of the Economies of Continental Europe, 1850–1914* (London: George Allen & Unwin, 1977).
Ministerium für Wirtschaft, Mittelstand und Verkehr, *Notwendige Massnahmen zur Verbesserung der Landesstruktur in NRW*, Part I: 'Analysis and Proposals for Improving Regional Structure' (Düsseldorf: 1964).
—— *Notwendige Massnahmen zur Verbesserung der Landesstruktur in NRW*, Part II: 'Structural Changes through New Political, Economic and Technical Developments' (Düsseldorf: 1966).
—— *Technologieprogramm-Energie* (Erste Fortschreibung) (Düsseldorf: July 1979).
—— *Energiebericht 82: Energiepolitik in Nordrhein-Westfalen* (Düsseldorf: June 1982).
MOMMSEN, H., PETZINA, D., and WEISBROD, B. (eds.), *Industrielles System und politische Entwicklung in der Weimarer Republik*, i and ii (Düsseldorf: Athäneum Verlag, 1974).
MONIG, W., et al., *Konzentration und Wettwerb in der Energiewirtschaft* (Munich: R. Oldenbourg Verlag, 1977).
MULLER, W., 'Die Neuordnung des Ruhrkohlenbergbaus', *Das Mitbestimmungsgespräch*, 2 (1969), 24.
MULLER-ARMACK, A., *Wirtschafslenkung und Markwirtschaft* (Hamburg: Verlag für Wirtschaft und Sozialpolitik, 1947).
MULLER-REISSMAN, K. F., 'Kriterien für die Energieversorgung: Herausforderung zum Vergleich', *Frankfurter Hefte*, 35 (1980).

NAPHTALI, F., *Wirtschaftsdemokratie* (Frankfurt: Europäische Verlagsanstalt, 1966; orig. published Berlin: Verlagsgesellschaft des Allgemeinen Deutschen Gewerkschaftsbundes, 1929).

National Coal Board et al., *Coal and Energy Policy in Europe* (London: National Coal Board, Dec. 1972).

NEULOH, O., *Der neue Betriebsstil* (Tübingen: J. C. B. Mohr (Paul Siebeck), 1960).

OBIJOU, K., 'Die Praxis von betrieblicher und überbetrieblicher Mitbestimmung im Bereich der Rurhkohle AG', *Mitbestimmungsgespräch*, 3 (1980).

OECD, *Energy R & D* (Paris: 1975).

OETTLE, K., 'Das öffentliche Unternehmen als Instrument staatlicher Politik', *Verwaltungswissenschaftliche Informationen*, 4 (1983).

Official Minutes of the North Rhine Westphalia Landtag Debates, Official Archive Copies, Jan. 1947–Aug. 1948.

OIDTMANN, P. H., 'Die technische und wirtschafliche Entwicklung des Aachener Steinkohlenbergbaus', i and ii, Ph.D. thesis, Rheinische-Westfälische Technische Hochschule, Aachen, 1955.

OLIVER, H. M., 'German Neoliberalism', *Quarterly Journal of Economics*, 14 (1960).

OLSEN, M. E. (ed.), *Power in Societies* (London: Collier Macmillan, 1970).

OTTO, B., *Gewerkschaftsbewegung in Deutschland* (Cologne: Bund-Verlag, 1975).

OWEN SMITH, E., *The West German Economy* (London: Croom Helm, 1983).

PARKER, W. N., 'Fuel Supply and Industrial Strength: A Study of the Conditions Governing the Output and Distribution of Ruhr Coal in the Late 1920s', Ph.D. thesis, Department of Economics, Harvard University, Cambridge, Mass., 1951.

——'Entrepreneurial Opportunities and Response in the German Economy', *Explorations in Entrepreneurial History*, 1 (Oct. 1954).

——'Entrepreneurship, Industrial Organisation and Economic Growth', *Journal of Economic History*, 14: 4 (Dec. 1954).

——and POUNDS, N. J. G., *Coal and Steel in Western Europe* (London: Faber & Faber, 1957).

PATERSON, W. E., and SMITH, G. (eds.), *The West German Model: Perspectives on a Stable State* (London: Frank Cass, 1981).

PEACOCK, A., et al., *The Economic Analysis of Government* (Oxford: Martin Robertson, 1979).

PEARCE, D. W., and ROSE, J. (eds.), *The Economics of Natural Resource Depletion* (London: Macmillan, 1975).

PETERS, H. R., *Grundzüge sektoraler Wirtschaftspolitik* (Stuttgart: Verlag Paul Herpt, 1971).

PLITZKO, A., 'Die Grundprobleme des Steinkohlenbergbaus', *Mitteilungen der List Gesellschaft*, 5: 10 (1966).

POTH, L., *Die Stellung des Steinkohlenbergbaus im Industrialisierungsprozeß* (Berlin: Duncker & Humblot, 1971).

POTTHOFF, E., *Der Kampf um die Mitbestimmung* (Cologne: Bund-Verlag, 1957).

POUNDS, N. J. G., *The Ruhr* (London: Faber & Faber, 1952).
PREECE, R. J. C., and TILFORD, R. B., *Federal Germany: Political and Social Order* (London: Oswald Wolff, 1970).
PRYM, A. M., *Staatswirtschaft und Privatunternehmung in der Geschichte des Ruhrkohlenbergbaus* (Essen: Verlag Glückauf, 1950).
RANFT, N., *Vom Object zum Subjekt* (Cologne: Bund-Verlag, 1988).
RAU, J., and JOCHIMSEN, R., *Die Zukunft der Kohle* (Düsseldorf: NRW-Economics Ministry, 1987).
RAY, G. F., and UHLMANN, L., *The Innovation Process in the Energy Industries* (London: Cambridge University Press, 1979).
REGUL, R., 'Wandlungen und Verflechtungen der Ruhrkohle', *Wirtschaftliche Nachrichten für Rhein und Ruhr*, 8: 33 (1927).
REINTGES, H., 'Grenzen der Rationalisierung im Steinkohlenbergbaus', *Glückauf*, 109: 7 (1973).
——'Neue Position der deutschen Steinkohle', *Glückauf*, 110: 23 (1974).
——'Warum Energiepolitik für die deutshe Steinkohle?', *Glückauf*, 112: 24 (1976).
——'Wie eine werantwortungsbewußte Energiepolitik aus heutiger Sicht aussehen muß', *Glückauf*, 113: 10 (1977).
——'Die Kohlesubventionen müssen verringert werden', *Frankfurter Allgemeine* (9 June 1990).
RIESSER, J., *Die deutschen Großbanken und ihre Konzentration* (Jena: Gustav Fischer, 1910).
RIMLINGER, G. V., 'National Differences in the Strike Propensity of Coal Miners: Experience in Four Countries', *International and Labour Relations Review*, 12: 3 (1959).
——*Welfare and Industrialization in Europe, America and Russia* (New York: John Wiley & Sons, 1971).
RINN, I., *Handbuch der Bergwirtschaft der Bundesrepublik Deutschland* (Essen: Verlag Glückauf, 1970).
RONGE, V., ' "Solidarische" Selbstorganistation der Wirtschaft—eine Alternative zur Politisierung im Spätkapitalismus', *Leviathan*, 2 (1978).
——(ed.), *Am Staat vorbei: Politik der Selbstregulierung von Kapital und Arbeit* (Frankfurt: Campus Verlag, 1980).
——and SCHMIEG, G., *Restriktionen politischer Planung* (Frankfurt: Athenäum Fischer Taschenbuch Verlag, 1973).
Ruhrkohle AG, *Annual Reports* (Essen: 1984/5/7/8/9).
RUPP, H. K., *Sozialismus und demokratische Erneuerung* (Cologne: Pahl-Rugenstein Verlag, 1974).
SACK, S., and WAWRZINEK, H., 'Charakter und Besonderheiten der Krise des westdeutschen Steinkohlenbergbaus', *Wirtschaftswissenschaften*, 17 (1969).
SCHAAF, P., *Ruhrbergbau und Sozialdemokratie* (Marburg: Verlag Arbeiterbewegung und Geisteswissenschaft, 1978).

SCHAAL, P., 'Möglichkeiten der Anpassung des Steinkohlenbergbaus an die veränderte Struktur des Energiemarktes', Ph.D. thesis, Albert-Ludwigs-Universität, Freiburg, 1961.

SCHIEDER, T., *The State and Society in Our Times* (Walton-on-Thames: Thomas Nelson, 1962).

SCHIRNER, K. (Generaldirektor, DKBL), *Kohle im Wirtschaftsaufbau: Aufbau der Kohlenwirtschaft*, Address at the Conference of Minister-Presidents (Düsseldorf: 5–6 June 1948).

Schmalenbach Commission, *Gutachten über die gegenwärtige Lage des Rheinisch-Westfälischen Steinkohlenbergbaus* (Berlin: Verlag Deutsche Kohlenzeitung, 1928).

SCHMIDT, A., *Lang war der Weg* (Bochum: IG Bergbau und Energie, 2nd edn., 1978).

SCHMIDT, C. D., 'Die Krise im Steinkohlenbergbau und ihre soziale Problematik unter besonderer Berücksichtigung des Ruhrgebietes', Ph.D. thesis, Westfälische Wilhelms-Universität, Münster, 1967.

SCHMIDT, H. (ed.), *Energiewirtschaft und Energiepolitik in Gegenwart und Zukunft* (Berlin: Duncker & Humblot, 1966).

SCHMOLLER, G., 'Das Verhältnis der Kartelle zum Staate', in *Schriften des Vereins für Sozialpolitik*, 116 (Leipzig: Duncker & Humblot, 1906).

SCHNEIDER, H., *Die Interessenverbände* (Munich: Gunter Olzog Verlag, 1965).

SCHNEIDER, H. K. 'Strukturwandlungen in der Energiewirtschaft der Bundesrepublik 1950–1960', in H. König (ed.), *Wandlungen der Wirtschaftsstruktur in der BRD* (Berlin: Schriften des Vereins für Sozialpolitik (N. F. 26), 1962).

SCHNEIDER, M., *Kleine Geschichte der Gewerkschaften* (Bonn: Verlag J. H. W. Dietz Nachf., 1989).

SCHUNDER, F., *Tradition und Fortschritt: Hundert Jahre Gemeinschaftsarbeit im Ruhrbergbau* (Stuttgart: W. Kohlhammer Verlag, 1959).

SEMRAU, G., 'Zur Wettbewerbssituation der inländischen Steinkohle gegenüber dem Öl', *Glückauf*, 117: 2 (1981).

SHEEHAN, J. J., *German Liberalism in the Nineteenth Century* (Chicago: University of Chicago Press, 1978).

SHONFIELD, A., *Modern Capitalism* (London: Cambridge University Press, 1965).

SHUCHMANN, A., *Co-determination: Labour's Middle Way in Germany* (Washington, DC: Public Affairs Press, 1957).

SIEBRECHT, F., 'Der Arbeitsdirektor im Bergbau', *Glückauf*, 1: 2 (1952).

SPD, *Die Kohle, Prof. Erhard und die SPD* (Bonn: SPD-Parliamentary Press Office, 1959).

SPD-Young Socialists (ed.), *Programme der deutschen Sozialdemokratie (1863–1963)* (Hanover: Verlag J. H. Dietz, 1963).

SPECHT, U., 'Die Energiepolitik der Bundesrepublik von 1948–1967', Ph.D. thesis, Albert-Ludwigs-Universität, Freiburg, 1969.

SPIEGELBERG, F., *Energiemarkt im Wandel: Zehn Jahre Kohlenkrise an der Ruhr* (Baden-Baden: Nomos Verlagsgesellschaft, 1970).

Statistik der Kohlenwirtschaft e. V., *Der Kohlenbergbau in der Bundesrepublik Deutschland im Jahre 1988* (Essen and Cologne: Sept. 1989).

STOCKDER, A. H., *Regulating an Industry: The Rhenish-Westphalian Coal Syndicate* (New York: Colombia University Press, 1932).

STOLPER, G., *The German Economy: 1870 to the Present* (London: Weidenfeld & Nicolson, 1967).

ST SEIDENFUß, H., 'Strukturwandlungen in der Energiewirtschaft', in F. Neumark (ed.), *Strukturwandlungen einer wachsenden Wirtschaft* (Berlin: Duncker & Humblot, 1964).

STUEBEL, H., *Staat und Banken im Preussischen Anleihewesen von 1871–1913* (Berlin: Carl Heymanns Verlag, 1935).

THIMM, A. L., *The False Promise of Co-determination* (Lexington, Mass.: Lexington Books, 1980).

THUM, H., *Mitbestimmung in der Montanindustrie: Der Mythos vom Sieg der Gewerkschaften* (Stuttgart: Deutscher Verlags-Anstalt, 1982).

THROM, W., 'Die Kohle und der Wettbewerb', in *Ordo*, xi (Düsseldorf: Helmut Kupper vormals Georg Bondi, 1959).

Transcripts of Motions submitted to the *Landtag* (NRW), Jan. 1947–Aug. 1948.

TSCHIRBS, R., 'Der Ruhrbergmann zwischen Privilegierung und Statusverlust: Lohnpolitik von der Inflation bis zur Rationalisierung (1919 bis 1927)', in G. Feldman *et al.* (eds.), *Die deutsche Inflation: Eine Zwischenbilanz* (Berlin: Walter de Gruyter, 1982).

UNGER, R. M., *Knowledge and Politics* (London: Macmillan, 1975).

Unternehmensverband Ruhrbergbau, *Das brennende Problem* (Essen: Verlag Glückauf, 1957).

Unterrichtung durch die Bundesregierung, *Die Energiepolitik der Bundesregierung*, Drucksache 7/1057 (1973) (Deutscher Bundestag 7. Wahlperiode).

——*Erste Fortschreibung des Energieprogramms der Bundesregierung*, Drucksache 7/2713 (1974) (Deutscher Bundestag 7. Wahlperiode). English trans. published by the Federal Minister of Economics, *First Revision of the Energy Policy Programme for the Federal Republic of Germany* (Nov. 1974).

VETTER, H. O., 'Zwanzig Jahre Mitbestimmungsgesetz-Geschichte oder Auftrag?', *Mitbestimmungsgespräch*, 5 (1971).

——*Mitbestimmung—Idee, Wege, Ziele* (Cologne: Bund-Verlag, 1979).

VILMAR, F., *Politik und Mitbestimmung* (Frankfurt: Athenäum Verlag, 1977).

——*Theorie der Wirtschaftsdemokratie*, Course of three lectures for the West German Open University (Fernuniversität), Hagen, 1980.

VON BECKERATH, G., 'Kostenstruktur und Marktordnung im Steinkohlenbergbau', *Zeitschrift für handelswirtschaftliche Forschung*, 8 (1956).

VON BENNIGSEN-FOERDER, R., *Energieversorgungsstrategien als unternehmerische Aufgabe*, Atomwirtschaft, Aug./Sept. 1982.

VON NELL-BREUNING, O., *Streit um die Mitbestimmung* (Frankfurt: Europäische Verlagsanstalt, 1968).

WALKER, F., 'Monopolistic Combinations in the German Coal Industry', *Publications of the American Economic Association*, 3rd ser. 5: 3 (Aug. 1904).
WATERKAMP, R., *Interventionsstaat und Planung* (Cologne: Verlag Wissenschaft und Politik, 1973).
WEGEHENKEL, P., *Die Beihilfen zur Bekäampfung der Steinkohlenkrisen in der Bundesrepublik Deutschland für den Zeitraum, 1958–1968* (Hamburg: Veröffentlichungen des HWWA-Institut für Wirtschaftsforschung Hamburg, 1970).
WELBERGEN, J. C., 'Die Zukunft der Kohle im Gesamtenergiebild aus der Sicht der Mineralölindustrie', in *Jahrbuch für Bergbau, Energie, Mineralöl und Chemie* (Essen: Verlag Glückauf, 1978/9).
WESSELS, T., 'Wirtschaftliche Probleme des Steinkohlenbergbaus in den letzten hundert Jahren', *Glückauf*, 95: 14 (1959).
WHITING, A. (ed.), *The Economics of Industrial Subsidies* (London: HMSO, 1976).
WILLECKE, R., *Die deutsche Berggesetzgebung* (Essen: Verlag Glückauf, 1977).
Wirtschaftsvereinigung Bergbau, *Soziale Marktwirtschaft auch für den Energiemarkt* (Essen: Verlag Glückauf, Oct. 1959).
WISOTSKY, K., *Der Ruhrbergbau in Dritten Reich* (Düsseldorf: Schwann, 1983).
WITT, W., 'Enqueten über den deutschen Steinkohlenbergbau und ihre Bedeutung für das Gesetzes- und Verordnungswesen', *Glückauf*, 100: 5 (1964).
WYN GRANT, E., *The Political Economy of Corporatism* (London: Macmillan, 1985).
'Zeit'/Diskussion, *Die Energiekrise* (Hamburg: Hoffmann und Campe, 1974).
ZIRANKA, J., 'Die Auswirkungen von Zechenstillegungen und Rationalisierungen im Steinkohlenbergbau auf die Wirtschaftsstruktur ausgewählter Gemeinden im niederrheinisch-westfälischen Industriegebiet', *Forschungsberichte des Landes NRW*, 1311 (1964).
ZWEIG, K., *Germany Through Inflation and Recession* (London: Centre for Policy Studies, 1976).
ZYDEK, H., 'Anpassungsgeld: Ein Beitrag zur Entstehungsgeschichte, zum Inhalt und zur Auslegung der Anpassungsgeldregelung', *Der Kompass*, 1 (1972).
ZYLSTRA, K., 'The Analysis of Structural Change: Energy Supplies', in J. R. Denton, and J. J. N. Cooper (eds.), *The European Economy Beyond the Crisis: From Stabilisation to Structural Change* (Bruges: De Tempel, 1977).